AI for Advanced Manufacturing and Industrial Applications

Bidyut Sarkar · Rudrendu Kumar Paul

AI for Advanced Manufacturing and Industrial Applications

Bidyut Sarkar
Edison, NJ, USA

Rudrendu Kumar Paul
San Francisco Bay, CA, USA

ISBN 978-3-031-86090-4   ISBN 978-3-031-86091-1 (eBook)
https://doi.org/10.1007/978-3-031-86091-1

This Springer imprint is published by the registered company Springer Nature Switzerland AG
The registered company address is: Gewerbestrasse 11, 6330 Cham, Switzerland

If disposing of this product, please recycle the paper.

Acknowledgements

This work owes much to the invaluable support of numerous individuals and organizations. The constructive feedback from the anonymous reviewers has significantly improved the manuscript's quality.

On a personal note, I am grateful to my family for their patience and understanding during the writing process. Their unwavering support made this work possible.

Bidyut Sarkar

I have been fortunate to have had the support and guidance of many talented mentors throughout my career. I am grateful to them for their wisdom and guidance and for helping me to develop the skills and knowledge needed to succeed in this field.

I would also like to thank my parents for their unwavering support and encouragement. Their belief in me has been a constant source of motivation and has played a crucial role in my professional development.

Finally, I would like to express my deep gratitude to my wife, whose love and support have been an invaluable source of strength and inspiration. Without her, this book would not have been possible.

Rudrendu Kumar Paul

Competing Interests The authors have no competing interests to declare that are relevant to the content of this manuscript.

Contents

About the Authors

Bidyut Sarkar is an IT professional specializing in the Life Science and Industrial Manufacturing industry. Over his 20-year career at IBM, he has developed a blend of industry experience, a comprehensive understanding of AI and Analytics, and strategic solutioning skills. His work has impacted large pharmaceutical companies and industrial manufacturing industries, particularly in the United States. He has been part of efforts to provide IT solutions against counterfeit drugs and implement AI/ML-powered predictive demand and automated replenishment capabilities. Through AI-driven technologies, he has contributed to solutions that enhance cyber-security and guarantee the authenticity and safety of medications worldwide. His professional experience has taken him to various parts of the world, including the USA, the Netherlands, Saudi Arabia, Brazil, and Australia. This multicultural experience has given him insights into global organizational challenges, reinforcing his commitment to addressing critical issues. He is an alumnus of IIT-Delhi, a premier institute in India.

Rudrendu Kumar Paul is an AI expert and applied ML industry professional with over 15 years of experience across multiple industries. He has several years of experience managing teams of data scientists and engineers. Rudrendu has a strong background in managing end-to-end machine learning processes at the top Fortune 50 companies in industrial applications, e-commerce, and high-tech indus-tries. His expertise includes advanced analytics, experimentation, customer lifetime value prediction, and deploying and monitoring machine learning models in produc-tion environments. Rudrendu's academic background includes an MBA, an MS in Data Science from Boston University (MA, USA), and a Bachelor's in Electrical Engineering.

List of Figures

List of Tables

Chapter 1
Introduction to AI in Manufacturing

Abstract The manufacturing industry is being redefined by Artificial Intelligence (AI) application by reshaping production processes, supply chain, and enterprise operations. This chapter provides a comprehensive introduction to AI in manufacturing, tracing its evolution from early automation in the 1970s to today's intelligent, interconnected systems. The chapter explores key milestones such as the integration of machine learning, Industry 4.0 advancements, and the rise of industrial IoT, which have enabled AI applications like predictive maintenance, quality control, and adaptive automation. The chapter discusses the significant market growth of AI, projected to reach $826.7 billion by 2030, alongside a detailed analysis of adoption trends, competitive dynamics, and the role of startups and technology leaders in driving innovation. Additionally, it addresses challenges like data accessibility, AI model transparency, and legacy system integration that have historically hindered adoption. Key AI applications enhancing manufacturing efficiency, quality, and sustainability are examined, alongside actionable insights into evaluating Return on Investment (ROI) through pilot projects and strategic metrics. The chapter provides a forward-looking perspective on AI's potential to further revolutionize manufacturing, positioning it as an essential tool for competitiveness and innovation in a data-driven industrial landscape.

Keywords Manufacturing ROI analysis · AI-driven manufacturing · Predictive maintenance · Intelligent quality control · Adaptive automation · Manufacturing data analytics

The manufacturing industry is on the cusp of an Artificial Intelligence (AI)-driven revolution, set to redefine manufacturing practices and pave the way for an era of efficiency and innovation. Interconnected systems, data-driven intelligence, and adaptive automation to reshape production, supply chain, and enterprise ecosystems.

AI is revolutionizing the manufacturing industry by optimizing production processes, enhancing supply chain efficiency, and enabling adaptive automation. AI technologies, such as predictive maintenance and intelligent quality control, are transforming traditional manufacturing practices into data-driven operations. For example, AI-powered predictive analytics can foresee equipment failures before they occur, reducing downtime and maintenance costs.

This chapter serves as a foundational guide to understanding the role of Artificial Intelligence (AI) in manufacturing. We will look at the industry's early automation days, explore the introduction of machine learning in factories, and discuss AI's pivotal role in optimizing supply chain. We will also examine how the unlocking of operational data through enhanced connectivity and the advent of Industry 4.0 have spurred exponential growth.

This chapter also sheds light on the market dynamics, presenting strong evidence of the AI market's significant growth, projected to hit US $826.70 billion by 2030. This upward trend is reflected in the manufacturing sector, with substantial growth anticipated across value, output, the number of enterprises, and employment levels. We'll also provide an overview of the competitive landscape, spotlighting the initiatives of key technology vendors and the existing market players and startups in the manufacturing sector in the development and adoption of AI solutions.

1.1 Brief History of AI Adoption in Manufacturing

The origins of automation in manufacturing date back to the 1970s with the introduction of robotics for welding and painting and the use of early computer-aided design (CAD) tools. However, these solutions offered narrow functionality, lacking the intelligent adaptation seen in today's systems. In the 1980s, expert systems, fuzzy logic controllers, and simple machine vision inspection capabilities demonstrated AI's potential to enhance quality and guide robotics. However, computing constraints and data accessibility issues limited the revolutionary impact.

The adoption of AI Bellwether Predictive analytics to optimize logistics planning proved an early precursor of widening AI applicability by the late 1990s. However, the lack of model transparency, small sample biases, and weak integration with enterprise platforms constrain scale. In the 2000s, the proliferation of Manufacturing Execution Systems (MES), Enterprise Resource Planning (ERP), and connectivity standards like OPC-UA through the 2000s paved the way for contextualized datasets. This foundation enabled the adoption of predictive maintenance and process quality algorithms.

In the early 2010s, the growth of industrial IoT, cheaper sensors, and cloud platforms provided unprecedented manufacturing data volume, opening the path for deep learning, digital twins, and reinforcement learning to interpret, simulate, and enhance multifaceted production environments holistically.

Throughout the previous decades, persistent challenges related to data accessibility, AI model transparency, and legacy environment integration have constrained widespread AI adoption in manufacturing. However, as factories began to overcome these hurdles and operational data became more abundant, the potential for AI integration grew rapidly.

The integration of AI and machine learning in manufacturing began to take root in earnest by 2012–2015, building upon the foundation laid by Industry 4.0 initiatives. This evolution was made possible by the convergence of several key factors: the increasing availability of big data from sensors and connected devices, advancements in computing power, and breakthroughs in machine learning algorithms. Early adopters started experimenting with AI for predictive maintenance and quality control during this period.

By 2016–2017, success stories from these pioneers sparked wider interest, leading to more widespread adoption. The pace of integration accelerated rapidly from 2018 onwards, with AI applications expanding into areas like adaptive manufacturing, supply chain optimization, and generative design. Today, AI and machine learning are no longer just experimental technologies in manufacturing but are becoming essential tools for companies looking to stay competitive in an increasingly digital and data-driven industry landscape.

These technologies are now transforming manufacturing in powerful ways. They help predict when machines need repairs, adjust production on the fly, and spot product defects more accurately than humans. AI also makes supply chain smarter and helps design better products, making factories more efficient, flexible, and able to create custom products faster. As AI continues to evolve, its integration into manufacturing processes promises to drive further advancements in efficiency, innovation, and sustainability.

1.2 Market Size and Growth Trends

The Artificial Intelligence (AI) market is on a trajectory for significant growth and a substantial increase in market size over the coming years. In 2024, the market is expected to reach a value of US$ 184.00 billion, and it is forecasted to maintain an annual growth rate (CAGR) of 28.46% from 2024 to 2030. The market is expected to grow to US $826.70 billion by the year 2030 [1].

During the same time, the manufacturing sector worldwide is expected to reach US $8.8 trillion in 2024, with a predicted compound annual growth rate (CAGR) of 1.15% over the next five years—marked by growth in value, output, enterprises, and employment [2].

The data presented indicates a strong growth trajectory for both the AI market and the Word Wide manufacturing sector. This growth is characterized by an impressive annual growth rate, reflecting the increasing importance and integration of AI technologies across various industries.

The manufacturing industry, broadly covering firms that transform raw materials and components into finished goods, is increasingly relying on AI to maintain high standards in production.

AI plays a key role in enhancing various aspects of production. According to a survey conducted by Bergur Thormundsson and published on March 17, 2022, a significant majority of respondents (59%) identified quality control as the most critical AI use case within the industry [3].

Quality control in manufacturing typically involves standardizing production processes to ensure consistency and quality of the final product. AI significantly contributes to this area by integrating advanced technologies like smart cameras, which streamline inspection processes. This integration not only improves the accuracy of quality control but also leads to cost reductions.

These trends signify a broader movement towards digital transformation in industries, with AI emerging as a key driver of efficiency, innovation, and competitive advantage. The data reflects the growing importance of AI in enhancing operational capabilities and shaping the future of industrial sectors.

AI has a growing and pivotal role in modernizing the manufacturing industry.

1.3 Competitive Landscape

The adoption of AI in manufacturing is evidence of a significant transformation. Existing market players are not only investing heavily in developing tailored machine-learning solutions for applications such as predictive maintenance and smart robotics but also leveraging cloud-based AI platforms. This strategic approach enables rapid prototyping, testing, and scaling of AI solutions.

Furthermore, the trend of significant mergers and acquisitions in this sector shows the industry's aggressive pursuit to integrate AI with existing products and services. This synergy between internal development and external collaboration is accelerating the democratization and adaptability of AI technologies in manufacturing, promising to unlock its vast potential across various production environments.

1.3.1 Vendor Offerings

Major manufacturing enterprises like GE, Siemens, Fanuc, Kuka, Bosch, and Honeywell are making substantial internal investments in developing customized machine-learning solutions suited for industry applications like predictive maintenance, production quality optimization, and smart robotics.

At the same time, these companies are strategically leveraging horizontal cloud-based AI development platforms and services from leading technology providers (such as IBM, Microsoft, AWS, and Google Cloud) to rapidly prototype, test, and scale up adoption.

For instance, Fanuc has invested in adding AI capabilities to its industrial robotic systems to enable more responsive, adaptive automation through real-time machine learning. Kuka offers a range of AI-enabled robotics leveraging cloud and edge computing innovations from partners like Microsoft Azure.

Key vendors offering AI software platforms, components, and manufacturing-focused capabilities helping drive adoption include

- IBM (Maximo industrial asset management, IBM Sterling supply chain)
- Microsoft (Dynamics 365 Connected Factory, Azure IoT, Machine Learning Services)
- Google Cloud (Vertex AI platform, AutoML Vision, Video & Image Annotations)
- AWS (Panorama appliance SDKs, Lookout for Equipment predictive maintenance, Monitron device monitoring)
- Siemens (MindSphere platform, PlantSight digital twin simulator)
- Oracle (OI Cloud Manufacturing Analytics, OCI GoldenGate for real-time data streaming).

These leading cloud and enterprise technology providers enable organizations to leverage AI/ML capabilities tailored to factory environments, reducing barriers to experimentation and integration.

This symbiotic approach allows manufacturers to concentrate resources around differentiating vertical capabilities while leveraging horizontal enablers to speed up deployment, adaptability, and democratization. The joint momentum promises to rapidly unlock AI's immense latent value across production environments.

1.3.2 Increased Mergers and Acquisitions (M&A) Activity

Recent high-profile mergers and acquisitions indicate that manufacturers plan to aggressively utilize AI and connectivity to gain market share. As AI and connectivity transform previously static products into continually self-optimizing intelligent systems, analysts expect marquee multi-billion-dollar acquisitions to accelerate.

For example, Rockwell Automation spent $2.6 billion acquiring Plex Systems—the cloud-based manufacturing analytics provider—in 2022 to combine its industrial automation expertise with Plex's AI-powered manufacturing intelligence [4].

Similarly, GE Healthcare acquired medical imaging AI specialist Zionexa for around $1 billion in 2022 to add machine learning-enhanced radiology and pathology solutions to its imaging portfolio. Zionexa's AI algorithms help detect the onset of chronic diseases years earlier.

1.4 Investor Landscape

The manufacturing sector is experiencing a significant shift, propelled by a surge of venture-backed startups specializing in applying AI to areas such as automotive, aerospace, and industrial equipment manufacturing. This surge is fueled by advancements in machine learning, computer vision, and natural language processing, which have expanded the range of AI applications in production environments [5].

The increasing availability of data from industrial IoT networks, coupled with the rise of cloud platforms, is accelerating the commercialization and scalability of AI solutions. Startups are focusing on critical areas such as predictive maintenance, quality control, and logistics optimization, signaling a future where AI becomes a foundational technology in large-scale manufacturing.

Expanding datasets from industrial internet-of-things sensor networks and equipment has helped improve model quality by providing relevant, high-fidelity data to train algorithms for manufacturing tasks. This unlocks enterprise-wide asset knowledge otherwise trapped in siloed operators into systematically improvable AI tools.

The growth of cloud platforms has simultaneously reduced infrastructure barriers to rapidly commercialize innovations at scale. Solutions built leveraging AWS, GCP, and Azure AI services can be quickly transitioned into production use by factories worldwide rather than remaining isolated prototypes.

These catalysts have spawned startups in the following manufacturing areas:

- Quality optimization
- Inventory and logistics planning
- Automated optical inspection
- Integration of intelligent capabilities into robotics
- Physics-based digital twin simulation to evaluate process improvements.

While economic shifts have resulted in some dampening of startup investment, marquee category leaders continue gaining the backing of manufacturing giants seeking to fund and eventually acquire innovators, filling gaps in their own solution stack. This offers alternative pathways to scale manufacturing AI adoption despite capital market fluctuations.

These early-stage companies are not only garnering attention through substantial funding rounds, partnerships, and early customer adoption but are also setting the stage for a broader, AI-driven revolution in manufacturing:

- Seeqc provides a digital quantum computing system on a chip. The company partners with BASF to explore applications of quantum computing in chemical reactions for industrial use [6].
- Optimitive is a SaaS company that boosts efficiency and sustainability in process industries like cement or chemicals by means of AI.
- Augury, an Israeli startup, provides predictive, prescriptive AI for industrial manufacturing. It uses purpose-built AI, trained by industry experts and a large data library, to help customers eliminate production downtime, improve process efficiency, maximize yield, and reduce waste and emissions [7].
- MakinaRocks is a startup specializing in industrial machine intelligence. It uses proprietary technology in anomaly detection and intelligent control [8].
- Landing AI, by Andrew Ng, provides a platform that allows companies to build, iterate, and deploy AI-powered visual inspection solutions for manufacturing [9].

- Osaro is a machine learning company that specializes in AI software for industrial automation, enabling retailers and logistics companies to deploy the systems faster.
- Drishti uses AI and computer vision to provide visibility and insights for manual assembly line improvement [10].
- OnScale offers a Solver-as-a-Service platform with advanced computer-aided engineering (CAE) multi-physics solvers that integrate with a scalable Cloud High-Performance Computing (HPC) platform [11].
- Covariant provides a universal AI (the Covariant Brain) to give robots the ability to see, reason, and act on the world around them [12].

As these AI technologies evolve from pilot projects to production-grade deployments, they demonstrate the potential for measurable and transformative impacts. Eventually, it's anticipated that these AI-based technologies will be adopted by the largest manufacturers, fundamentally altering the landscape of the manufacturing sector.

1.5 Evaluating Return on Investment (ROI) for AI Projects

Embracing AI isn't merely about being in vogue with the latest technological trends. It's about extracting tangible, actionable value. Implementing AI projects often demands a significant outlay, both in terms of capital and time. Yet, the returns—whether they manifest as cost savings, time efficiencies, or broader strategic leverage—can position a company distinctly ahead in the competitive arena.

For AI projects, this means that a comprehensive evaluation of the Return on Investment (ROI) is necessary. Assessing ROI for AI projects demands a meticulous analysis spanning multiple financial and strategic metrics, ranging from immediate, measurable gains to long-term strategic advantages.

A survey research data on the global return on investment (ROI) for artificial intelligence (AI) from 2015 to 2019, specifically focusing on the first year of AI deployment in various companies. Over this period, approximately 1000 professionals from multiple industries were surveyed regarding the ROI delivered by AI in their organizations. The findings indicate that the ROI from AI either met or exceeded initial expectations [13].

AI's predictive analytics capabilities have not only helped manufacturers to undertake more informed decision-making but also chart the course for strategic planning. Be it foretelling customer behavior, anticipating market fluxes, or preempting system failures, AI's predictive analytics emerge as an indispensable strategic asset for organizations.

Figure 1.1 provides a visual representation of the financial trajectory of an AI project, capturing the dynamic interplay between cost, revenue, and return on investment (ROI) over time. The figure illustrates the initial investments required for an AI initiative and the long-term financial benefits that can be realized.

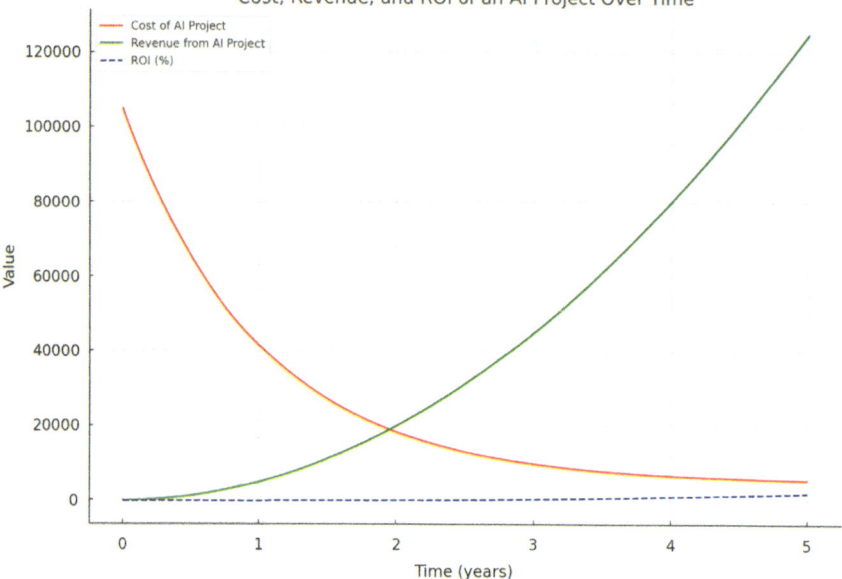

Fig. 1.1 Variation of cost, revenue, and ROI for an AI project, both initial investments and long-term gains. Data is for illustrative purposes only. *Source* Created by the authors in Python Notebook

In the above figure,

- The red curve represents the decreasing cost of the AI project over time. This decrease can be attributed to the amortization of initial investments and the realization of efficiencies from the AI system.
- The green curve shows the increase in revenue as the AI system becomes more integrated and optimized and starts delivering tangible results.
- The blue dashed curve represents the ROI for AI projects. As evident, the ROI improves over time, indicating that the AI project becomes more profitable.

The return on investment (ROI) for AI projects is calculated using the formula

$$ROI = (Revenue - Cost)/Cost \times 100$$

This formula calculates the ROI as a percentage. The result indicates the return on the investment relative to its cost. A positive ROI indicates that the investment has gained value, while a negative ROI indicates a loss.

To effectively evaluate the ROI of AI projects, one must consider a multi-faceted approach. This begins with understanding immediate, measurable metrics such as time reduction, cost savings, and conversion rates. However, it doesn't stop there. For a holistic view, one must also look at the strategic ROI, to ensure that AI projects align with the organization's long-term goals and market positioning.

By combining these perspectives, organizations can paint a comprehensive picture of the true value AI brings, ensuring that their investments are sound and future-proof.

1.5.1 Important Financial Metrics

Financial metrics serve as quantifiable measures that provide a clear picture of an organization's performance, especially when integrating new technologies like AI. For AI projects, some of the pivotal metrics include improving process efficiency, cost reduction, conversion rate to sale, improved customer lifetime value, and risk mitigation.

Some of the important metrics are given in Table 1.1

To effectively leverage these metrics, one should:

Identify the appropriate Benchmarks

Before implementing AI, establish a clear baseline for each metric to measure against post-implementation. This will help to evaluate the impact and value of AI on the organization's performance and goals. For example, a manufacturing company can

Table 1.1 Main financial metrics

Financial metric	Description	Example
Process efficiency improvement	Measures how effectively AI automates and streamlines operations, resulting in time and cost savings	A company implemented AI for data processing, reducing processing time by 30% and manual effort
Operational cost reduction	Evaluates the reduction in costs achieved through AI-driven automation, error reduction, and process optimization	Automating customer service with an AI chatbot led to a 25% decrease in support costs
Increased conversion rates	Tracks the boost in sales conversion due to AI-powered personalized recommendations and targeted analytics	An e-commerce platform saw its conversion rate rise from 2 to 4% after adopting AI for product suggestions
Enhanced customer lifetime value	Assesses the growth in the long-term value of customers, supported by AI-driven personalization and customer insights	A subscription service used AI to offer tailored content, resulting in a 20% increase in customer retention
Risk mitigation	Measures AI's capability to identify and mitigate potential risks, including operational, financial, or cybersecurity threats	An AI system proactively detected security vulnerabilities, reducing potential breach risks by 40%

measure its current production time, quality, and costs before using AI to enhance its processes and products.

Monitor Continuously

AI's impact is often dynamic and evolving. Regularly track and analyze these metrics to understand AI's ongoing value and to recalibrate strategies as needed. This will help to optimize AI's performance and alignment with the organization's vision and mission.

For example, an automobile manufacturing company monitored its customer service metrics, such as response time, resolution rate, and feedback score, to assess and improve its AI chatbot's functionality and quality.

1.5.2   Executive Guide to AI Project ROI Analysis

This section discusses the three fundamental dimensions of AI Return on Investment (ROI)—Measurable, Strategic, and Capability ROI (Fig. 1.2).

For each dimension, we'll discuss how AI investments can drive not just immediate financial gains but also long-term strategic benefits and enhanced organizational capabilities in AI.

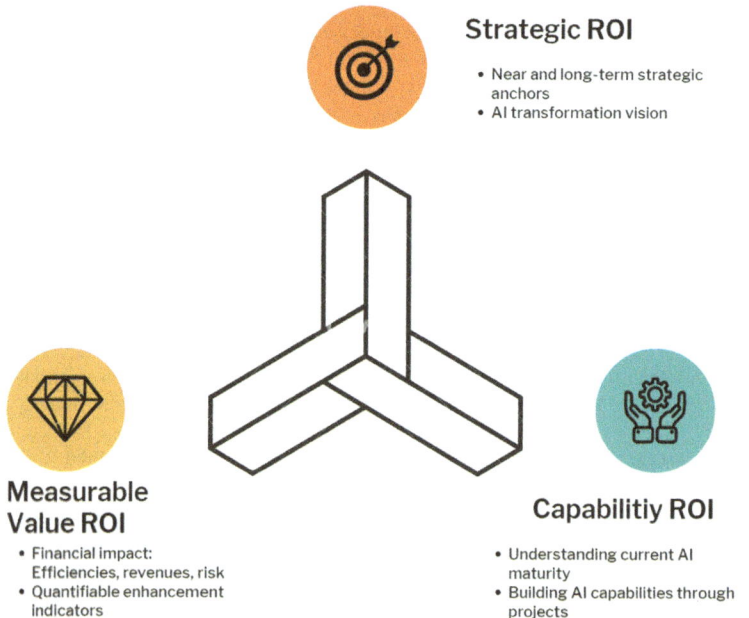

Strategic ROI
- Near and long-term strategic anchors
- AI transformation vision

Measurable Value ROI
- Financial impact: Efficiencies, revenues, risk
- Quantifiable enhancement indicators

Capabilitiy ROI
- Understanding current AI maturity
- Building AI capabilities through projects

Fig. 1.2 Executive framework for analyzing AI project ROI across three dimensions—measurable, strategic, and capability ROI. *Source* Created by the authors in Canva

Measurable ROI

Measurable ROI focuses on the quantifiable impacts of AI projects. It encompasses aspects like financial gains through cost savings or increased revenue, as well as non-financial gains like improved customer satisfaction or operational efficiencies. For instance, automating processes can save time and resources, directly impacting the bottom line. Enhanced revenues can be achieved through strategies like personalized marketing or optimized product recommendations.

Strategic ROI

Strategic ROI addresses AI's alignment with a company's long-term goals and strategic vision. This aspect involves identifying how AI initiatives support key business objectives, such as gaining market share or enhancing customer experience. The process entails filtering AI projects through strategic goals to ascertain their relevance and impact.

Capability ROI

Capability ROI is concerned with the long-term impact of AI projects on a firm's overall AI maturity. It's not just about immediate project outcomes but about building a foundation for future AI innovation. This involves cultivating an AI-savvy culture, acquiring necessary skills, and developing suitable infrastructure. AI projects should be viewed as investments in a firm's capability to utilize AI effectively in the market.

The goal is to transition from viewing AI as just another IT tool to recognizing it as a strategic asset that requires a unique approach in terms of data handling, team collaboration, and continuous learning and adaptation.

Embracing these dimensions holistically ensures that AI investments are not only financially sound but also strategically aligned and future-proof, positioning firms to thrive in an increasingly AI-centric business environment.

1.5.3 Pilot AI Projects

Pilot projects serve as preliminary tests or trial runs for larger AI implementations. They provide a controlled environment to assess the feasibility and impact of AI solutions. One can measure the efficiency, productivity, and profitability of a pilot project and compare it to other alternatives. To gauge the success of these pilot projects, it's essential to calculate specific financial metrics that reflect the project's performance and potential scalability.

Before scaling up, organizations need to identify any potential issues or challenges. By assessing the financial outcomes of a pilot, they can determine if the AI solution meets expectations or if adjustments are needed. For example, they can calculate the return on investment (ROI), the cost–benefit ratio, the break-even point, and the payback period of the pilot project.

The organization can also estimate the resource requirements and availability for scaling up the pilot AI solution. These metrics can help optimize resource allocation and ensure the best use of assets and capabilities.

Demonstrating positive financial outcomes from pilot projects can boost stakeholder confidence, ensuring continued support and investment in AI initiatives. The outcomes can showcase the value added, the customer satisfaction, the competitive advantage, and the social impact of the pilot project.

They can also highlight the potential opportunities and challenges for scaling up the AI solution. These metrics can help them communicate the vision and the benefits of the AI solution and persuade the stakeholders to back it up.

The results from pilot projects can inform broader strategic decisions, helping organizations plan their next steps with clarity and confidence. For example, they can analyze the market demand, customer feedback, industry trends, and the regulatory environment of the pilot project. They can also explore the scalability, feasibility, sustainability, and innovation potential of the AI solution.

To effectively calculate and showcase the financial metrics of pilot AI projects:

Define Clear Objectives

Before initiating a pilot project, it's crucial to define clear financial objectives. These could include specific targets for cost reduction, efficiency improvement, or revenue increase. Having clear goals from the outset provides a benchmark against which the project's success can be measured.

Doing this also ensures alignment between the project outcomes and the organization's strategic goals. Clear objectives guide the project's direction and help stakeholders understand what the project aims to achieve.

Gathering Data

Data collection is a continuous and critical process throughout the duration of the pilot project. This includes both financial data and other relevant performance indicators. The data gathered can provide insights into the project's progress and the effectiveness of the AI solution. It's important to ensure that the data collected is accurate, relevant, and consistent to support reliable analysis and conclusions.

Analyze and Compare

Once the pilot concludes, it's important to analyze the data and compare the results against the predefined objectives. This analysis can reveal whether the AI solution met, exceeded, or fell short of the expectations. It can also identify trends, patterns, and anomalies that could provide valuable insights. The comparison helps in understanding the gaps, if any, and in planning necessary adjustments for future implementations.

Present Findings

After the analysis, the findings should be consolidated into a comprehensive report or presentation. This report should highlight key financial metrics, discuss the implications for the broader organization, and provide recommendations for future action.

The presentation of findings is a crucial step in communicating the value and impact of the AI solution to stakeholders. It helps build consensus, secure buy-in, and inform decision-making for future AI initiatives.

For organizations embarking on AI projects, estimating ROI becomes a critical step in the planning and execution process. It not only provides a clear financial roadmap but also ensures that the project aligns with the organization's broader strategic goals.

1.6 Checklist for Evaluating Return on Investment (ROI) for AI Projects

1. Define Clear Objectives

   Set Financial Targets: Establish specific targets for cost reduction, efficiency improvement, or revenue increase.
   Align with Strategic Goals: Ensure objectives align with the organization's broader strategic vision.

2. Identify Key Financial Metrics

   Improve Process Efficiency: Measure time savings and automation benefits.
   Cost Reduction: Assess decreases in operational costs due to AI implementation.
   Conversion Rate to Sale: Track increases in sales conversions from AI-driven recommendations.
   Improved Customer Lifetime Value: Evaluate how AI enhances customer retention and value.
   Reducing Risk: Measure AI's impact on mitigating operational, financial, or cybersecurity risks.

3. Establish Baselines

   Benchmark Current Performance: Measure existing performance metrics before AI implementation.
   Document Initial Costs: Record all upfront investments required for the AI project.

4. Pilot AI Projects

   Set Pilot Objectives: Define clear, measurable goals for the pilot project.
   Gather Data Continuously: Collect financial and performance data throughout the pilot.
   Analyze Results: Compare pilot outcomes against predefined objectives.
   Present Findings: Consolidate findings into a report, highlighting key metrics and insights.

5. Data Collection and Analysis

   Ensure Data Accuracy: Collect accurate, relevant, and consistent data.
   Use Advanced Analytics: Apply AI and ML techniques for data analysis.
   Identify Trends and Anomalies: Look for patterns and outliers that provide insights into AI performance.

6. Monitor Continuously

   Track Ongoing Metrics: Regularly monitor and analyze key performance indicators.
   Adjust Strategies as Needed: Recalibrate strategies based on real-time data and insights.

7. Evaluate Long-term Strategic ROI

   Assess Strategic Alignment: Ensure AI projects support key business objectives.
   Measure Capability ROI: Evaluate the long-term impact on organizational AI maturity and infrastructure.

8. Calculate ROI

   Use the ROI Formula: ROI = (Revenue – Cost) / Cost x 100
   Compare Short-term and Long-term Gains: Assess both immediate financial gains and long-term benefits.

9. Gather Stakeholder Feedback

   Engage Key Stakeholders: Collect feedback from all relevant parties to gauge satisfaction and gather insights.
   Adjust Based on Feedback: Use stakeholder feedback to refine AI strategies and implementations.

10. Communicate Results

   Prepare Reports and Presentations: Highlight key findings, metrics, and recommendations.
   Demonstrate Value: Showcase the financial and strategic value of AI investments to stakeholders.
   Plan Next Steps: Use insights from the evaluation to inform future AI initiatives and scale successful pilots.

1.7 Key Takeaways

- AI is transforming the manufacturing industry by enhancing efficiency, innovation, and interconnected systems, leading to a substantial reshaping of production and supply chain.

- The transition from basic automation using robotics in the 1970s to sophisticated AI applications today highlights the evolutionary journey of AI in manufacturing.
- The AI market is projected to grow significantly, reaching US $826.70 billion by 2030, with the manufacturing sector also showing robust growth trends.
- AI's role in quality control is pivotal, improving inspection processes and operational consistency, which in turn drives cost efficiency.
- Major manufacturing firms and tech giants are increasingly investing in AI to develop customized solutions and leveraging cloud-based platforms for rapid testing and scaling.
- Startups are crucial in driving innovation in AI applications in manufacturing, with many focusing on predictive maintenance, quality optimization, and logistics planning.
- Evaluating the ROI of AI projects is essential, covering measurable gains like cost savings, strategic alignment with long-term goals, and building capabilities for future AI use.
- Conducting pilot AI projects helps assess the feasibility, measure success through specific financial metrics, and identify areas for improvement before wider implementation.
- The effective use of data, including collection, analysis, and interpretation, is central to maximizing the benefits of AI in manufacturing, enabling better decision-making and strategy development.
- As AI technologies continue to advance, their integration into manufacturing processes is expected to deepen, leading to more automated, efficient, and data-driven production environments.

Glossary

AI (Artificial Intelligence) Technology that enables machines to simulate human intelligence, including learning, problem-solving, and decision-making.

Cloud Computing The delivery of computing services over the internet, including servers, storage, databases, networking, software, and analytics.

Computer Vision AI technology that enables machines to interpret and understand visual information from the world.

Deep Learning An advanced form of machine learning based on artificial neural networks.

Digital Twin A virtual representation of a physical object or system used for simulation and optimization.

ERP (Enterprise Resource Planning) Software that manages and integrates a company's financials, supply chain, operations, and human resource activities.

Industry 4.0 The fourth industrial revolution, characterized by the integration of digital technologies, AI, and IoT in manufacturing processes.

IoT (Internet of Things) A network of interconnected devices that collect and exchange data.

Machine Learning A subset of AI that enables systems to learn and improve from experience without explicit programming.

MES (Manufacturing Execution System) Computerized systems are used to track and document the transformation of raw materials into finished goods.

Natural Language Processing (NLP) AI technology that enables machines to understand, interpret, and generate human language.

Predictive Analytics The use of data, statistical algorithms, and machine learning techniques to identify the likelihood of future outcomes based on historical data.

Predictive Maintenance Using data analysis to predict when equipment might fail and perform maintenance proactively.

ROI (Return on Investment) A performance measure used to evaluate the efficiency or profitability of an investment.

References

1. Statista. Artificial Intelligence—Global | Statista Market Forecast [Internet]. *Statista*. 2024. Access date November 21, 2024. https://www.statista.com/outlook/tmo/artificial-intelligence/worldwide
2. Statista. Manufacturing—Worldwide | Statista Market Forecast [Internet]. *Statista*. 2024. Access date November 21, 2024. https://www.statista.com/outlook/io/manufacturing/worldwide1
3. Statista.com. AI usage manufacturing industry worldwide 2020 [Internet]. *Statista*. 2022. Access date November 21, 2024. https://www.statista.com/statistics/1197949/ai-manufacturing-industry-use-case-worldwide/
4. Rockwell completes its acquisition of Plex systems | Rockwell Automation [Internet]. *Rockwell Automation*. 2023. Access date November 21, 2024. https://www.rockwellautomation.com/en-cn/company/news/press-releases/rockwell-completes-its-acquisition-of-plex-systems.html
5. Newsroom | GE Healthcare [Internet]. *Gehealthcare.com*. Access date November 21, 2024. https://www.gehealthcare.com/about/newsroom/press-releases/ge-healthcare-acquires-zionexa-molecular-imaging-agent-aims-to-enable-more-0?npclid=botnpclid
6. SEEQC partners with BASF to explore applications of quantum computing in chemical reactions for industrial use [Internet]. *SEEQC*. 2024. Access date November 21, 2024. https://seeqc.com/press/seeqc-partners-with-basf-to-explore-applications-of-quantum-computing-in-chemical-reactions-for-industrial-use
7. Predictive, prescriptive AI for industrial manufacturing [Internet]. *Augury*. Access date November 21, 2024. https://www.augury.com/
8. Crunchbase [Internet]. *Crunchbase*. 2017. Access date November 21, 2024. https://www.crunchbase.com/organization/makinarocks
9. About Landing AI [Internet]. *LandingAI*. Access date November 21, 2024. https://landing.ai/about-us
10. Jensen, S. (2022). Drishti combines AI and video to improve manufacturing operations [Internet]. Powermotiontech.com. *Power & Motion*. Access date November 21, 2024. https://www.powermotiontech.com/video/video/21240963/drishti-drishti-combines-ai-and-video-to-improve-manufacturing-operations

11. OnScale exits stealth with $3M in seed funding and unveils first solver-as-a-service platform [Internet]. *HPCwire*. 2018. Access date November 21, 2024. https://www.hpcwire.com/off-the-wire/onscale-exits-stealth-3m-seed-funding-unveils-first-solver-service-platform/

12. AI Startup Covariant.ai Building "Universal AI for Robots" | Synced [Internet]. Synced | AI Technology & Industry Review. 2020. Access date November 21, 2024. https://syncedreview.com/2020/06/16/ai-startup-covariant-ai-building-universal-ai-for-robots/

13. AI return on investment worldwide 2019 [Internet]. *Statista*. Access date November 21, 2024. https://www.statista.com/statistics/1197801/ai-roi-worldwide/

Chapter 2
AI-Driven Manufacturing Processes

Abstract This chapter explores the use cases and applications of AI in manufacturing, focusing on how AI-driven technologies are transforming industrial processes, enhancing efficiency, and improving product quality. Predictive maintenance emerges as a cornerstone application, leveraging machine learning to analyze time-series sensor data and enabling condition-based maintenance over traditional time-based models. These advancements reduce equipment downtime, extend machine lifespan, and foster operational reliability. AI-powered computer vision revolutionizes quality assurance by automating defect detection, part inspections, and dimensional monitoring, significantly improving accuracy and speed. The chapter also examines the role of natural language processing (NLP) in extracting actionable insights from unstructured text data, such as maintenance logs and technician notes, streamlining operations and supporting predictive maintenance. Advanced anomaly detection systems employ AI to identify irregularities in real-time, ensuring consistent product quality and operational efficiency. The integration of ARM architecture and digital twin simulations enhances computational efficiency, enabling real-time process monitoring, virtual scenario testing, and optimization. Digital twins, powered by real-time sensor data and AI, offer dynamic simulations to identify bottlenecks and proactively implement improvements. This chapter provides a comprehensive overview of benefits of AI application in manufacturing and its potential to revolutionize processes, reduce costs, and drive innovation, making it indispensable for modern industrial applications.

Keywords Predictive maintenance · AI-driven quality control · Digital twin simulations · NLP in operations and maintenance · AI-based anomaly detection systems · ARM architecture in manufacturing

This chapter explores the impact of AI-driven manufacturing processes, offering key insights into how artificial intelligence is reshaping the landscape of industrial operations. The first section highlights the role of predictive maintenance, utilizing machine learning (ML) to analyze sensor data for forecasting equipment failures. This approach marks a significant shift from traditional time-based maintenance to a more efficient condition-based model, leveraging AI for pattern identification in time-series data and managing downtime through the prediction of potential failures.

The shift in maintenance strategies highlights how AI enhances operational reliability and helps prevent unexpected equipment failures. The following sections discuss about AI applications in manufacturing. For example, computer vision techniques, powered by deep learning, automate tasks such as visual defect detection, part inspection, and quality control, effectively identifying surface defects and monitoring product dimensions with increased accuracy. Additionally, natural language processing (NLP) extracts valuable insights from unstructured text data found in maintenance logs, technician notes, and other documentation, enabling the analysis of recurring issues and optimizing support processes.

Additionally, the chapter discusses the integration of ARM architecture in smart manufacturing environments for enhanced efficiency and the use of digital twin simulations. These simulations, powered by AI, create virtual models of systems and processes, enabling real-time data integration and operational testing in virtual scenarios. Together, these sections provide a comprehensive overview of how AI is not just automating but intelligently revolutionizing manufacturing processes.

2.1 Predictive Maintenance to Analyze Sensor Data and Forecast Equipment Failures

According to Grand View Research, the global market for predictive maintenance is set for remarkable growth. Valued at $9.84 billion in 2023, it is projected to soar to an impressive 60.13 billion U.S. dollars by 2030. Here, we will explore how machine learning leverages sensor data to anticipate equipment failures, offering a proactive approach to maintenance and a move towards adopting predictive technologies in modern industries [1].

Predictive maintenance has applications across multiple use cases for modern manufacturing through the use of Machine Learning (ML) to revolutionize how equipment health is monitored and managed. By analyzing vast streams of sensor data, ML offers a proactive approach, allowing manufacturers to anticipate equipment failures before they even occur. This shift is not just about preventing unplanned downtimes; it's about optimizing operational efficiency, reducing costs, and ensuring a seamless production process.

Through the lens of ML, we transition from a reactive stance, where issues are addressed post-failure, to a proactive paradigm, where potential problems are identified and rectified in advance. This section will discuss the intricacies of ML-driven predictive maintenance and how it can set new standards in manufacturing reliability and excellence.

Many companies currently view predictive maintenance as the prime method for utilizing industrial data analytics. Predictive maintenance can decrease equipment downtime and extend machine lifespan significantly.

2.1.1   Integration of IoT Sensors with AI for Enhanced Predictive Maintenance

Integrating IoT, analytics, and AI can be utilized to develop an advanced predictive maintenance framework. IoT devices serve as the "sensory organs," collecting real-time data from various machines and equipment. This data, however, can be vast and complex, making it difficult to process and analyze using traditional methods. This is where AI and analytics step in [2].

AI algorithms can learn from data and predict potential failures and maintenance needs long before they occur. These predictions are based on the continuous analysis of IoT device data streams, including variables such as temperature, vibration, and operational metrics. Combining IoT's data-gathering prowess with AI's analytical capabilities makes predictive maintenance strategies not only possible but highly efficient.

One of the most significant challenges in managing the enormous volumes of data produced by IoT devices is making sense of it. AI excels in this context. Through machine learning and deep learning techniques, AI can uncover patterns and anomalies in data that would otherwise go unnoticed.

For instance, AI can identify a subtle change in the vibration pattern of a machine, signaling a potential fault. By providing context to these data points, AI enables maintenance teams to understand the 'why' and 'how' behind potential issues, leading to more targeted and effective maintenance actions.

Analytics and data visualization are crucial in translating complex data sets into actionable insights. Advanced analytics tools can process and analyze IoT data, presenting it in a user-friendly format. Data visualization helps simplify this information through graphs, charts, and dashboards, making it accessible even to those without a technical background. This aspect of the technology allows maintenance teams to identify trends, patterns, and anomalies quickly. For example, a sudden temperature or energy usage spike can be easily spotted on a dashboard, prompting immediate action.

2.1.2   Case Study for Real-World Predictive Maintenance

Several real-world predictive maintenance case studies discussed below demonstrate significant cost savings and operational improvements across various industries that can be achieved using predictive maintenance [3].

Case Study 1: Oil and Gas Industry

In the oil and gas industry, predictive maintenance has been key in preventing equipment failures and reducing downtime. For instance, a leading oil and gas company implemented predictive maintenance using FAT FINGER's digital workflow builder. The company was able to anticipate equipment failures, schedule maintenance tasks efficiently, and significantly reduce downtime. This resulted in substantial cost savings and improved operational efficiency.

Case Study 2: Manufacturing Industry

In the manufacturing sector, a renowned automobile manufacturer used predictive maintenance to optimize its production line. By integrating FAT FINGER's predictive maintenance workflows, the company was able to predict potential equipment failures and schedule maintenance tasks accordingly. This proactive approach reduced unexpected breakdowns, improved production efficiency, and resulted in significant cost savings.

Case Study 3: Energy Sector

In the energy sector, a leading power generation company used predictive maintenance to enhance the reliability and efficiency of its power plants. By leveraging FAT FINGER's AI-powered predictive maintenance workflows, the company was able to predict potential equipment failures, schedule maintenance tasks proactively, and significantly reduce downtime. This resulted in improved power plant reliability, efficiency, and substantial cost savings.

2.1.3 Identification of Patterns Using ML Models in Time-Series Data for Predicting

In modern manufacturing, the ability to predict and prevent equipment failures is paramount. At the heart of this capability lies the identification of patterns using machine learning (ML) models, specifically within time-series data.

Time-series data, a sequence of data points indexed in time order, offers a chronological account of machine operations. By analyzing this data, ML models can discern patterns that may indicate impending failures.

2.1.4 Example Use Case

Reducing Equipment Downtime in Manufacturing with ML-Powered Predictive Maintenance in the illustrative example

At a manufacturing company, a team of data scientists deployed a Machine Learning model to scrutinize time-series data for predictive maintenance. The time series data for the machine operations provided invaluable insights for the model. Through rigorous model training, the team enabled the model to detect patterns signaling potential equipment failures. Transitioning from conventional methods, the company embraced this ML-driven approach, which remarkably cut downtime. Predictive maintenance can decrease equipment downtime by 30–50% and extend machine lifespan by 20–40%. This advancement in predictive maintenance not only increased manufacturing efficiency but also significantly curtailed unforeseen equipment downtime. The figure below demonstrates the efficacy of this approach, showcasing the impact of ML-enabled predictive maintenance on equipment performance metrics.

Figure 2.1 illustrates the significant impact of machine learning-powered predictive maintenance on various manufacturing equipment metrics.

Consider the vast amount of data generated by manufacturing equipment. Each piece of data, from temperature fluctuations to on/off cycles, tells a story. When these individual stories are pieced together, they can reveal a larger narrative about the health and efficiency of the machinery.

For instance, a sudden spike in temperature readings might be an isolated incident, or it could be a precursor to a significant malfunction. By identifying such patterns, manufacturers can preemptively address issues, leading to reduced downtime, increased machine longevity, and significant cost savings.

A prime example can be drawn from a study conducted on the auxiliary polymer manufacturing process. Here, multivariate time series data was used to automatically pinpoint recurring events, such as failure patterns, in real-world manufacturing data. The study revolved around the drying and feeding of plastic granulates to machines. By employing selected data mining techniques, the research identified unique patterns in the data set.

One approach used heuristic segmentation and clustering, while another incorporated a method with a built-in time dependency structure. The insights gleaned from

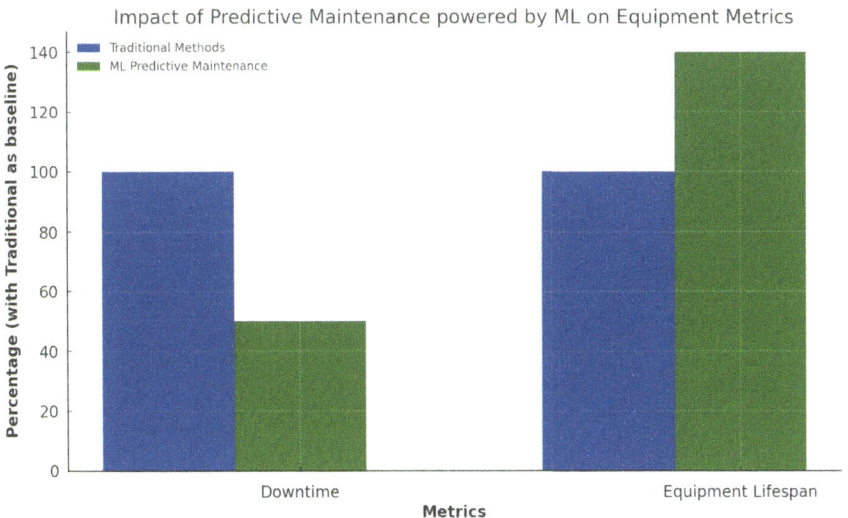

Fig. 2.1 Impact of predictive maintenance powered by ML on manufacturing equipment metrics. Data is for illustrative purposes only. *Source* Created by the authors in Python Notebook

Table 2.1 Overview of predictive maintenance: applications of ML for proactive equipment health monitoring

Key topic	Description	Impact of AI applications
Data source	Sensor data in time-series format	A chronological account of machine operations, capturing intricate details of every operational facet
Approach	Machine Learning (ML) analysis	Transforms vast streams of sensor data into actionable insights by recognizing patterns indicative of future failures
Operational shift	From reactive to proactive	Instead of addressing issues post-failure, predictive maintenance identifies and rectifies potential problems in advance
Key metrics	Downtime and equipment lifespan	Predictive maintenance can decrease equipment downtime by 30–50% and extend machine lifespan by 20–40%
Case study	Auxiliary polymer manufacturing process	Multivariate time series data was used to identify recurring failure patterns in real-world manufacturing data, specifically in the drying and feeding of plastic granulates
Practical implication	Equipment with sensors	In a setting where machinery is equipped with sensors, such as an industrial hopper drying plastic granulates, recurring patterns (e.g., temperature drops) can signal equipment flaws

these patterns were then applied to develop advanced monitoring methods, aiming to predict and mitigate future machine failures (Table 2.1).

In practical terms, this could be visualized in a manufacturing setting where plastic granulates are dried in an industrial hopper equipped with sensors. These sensors, constantly relaying data, might detect a recurring drop in temperature every few hours. This pattern, once identified, could indicate a flaw in the heating mechanism, prompting immediate action.

The identification of patterns via ML models in time-series data can be utilized for predictive maintenance. By understanding the significance of these patterns and creating ML models to detect these patterns, industries can upgrade their approach to equipment maintenance and set the stage for a more efficient and cost-effective manufacturing future.

2.1.5 Summary

Predictive maintenance with ML transforms manufacturing by preventing issues before they arise, enhancing efficiency, and reducing costs.

- Predictive maintenance, powered by ML, shifts the manufacturing approach from reactive problem-solving to proactive issue prevention, drastically reducing unplanned downtimes.

- By analyzing time-series data, ML can identify patterns indicative of potential equipment failures, enabling timely interventions and ensuring seamless production processes.
- Early detection and rectification of potential issues through predictive maintenance can extend machine lifespan and reduce maintenance costs, leading to substantial savings for manufacturers.

2.1.6 AI-Enabled Transition from Time Based to Condition-Based Maintenance

The integration of AI with condition-based monitoring (CBM) can be adopted for modernizing maintenance practices in the manufacturing industry. At its core, CBM focuses on real-time monitoring of equipment conditions to determine when maintenance interventions are necessary. This contrasts with traditional time-based maintenance, where interventions are scheduled at regular intervals, regardless of the actual condition of the equipment.

2.1.6.1 Example Use Case

Shifting to AI-Enabled Condition-Based Monitoring in Equipment Maintenance in the illustrative example

At a manufacturing facility, an initiative was launched to transition from time-based to AI-enabled condition-based maintenance. This approach harnessed AI algorithms to continuously monitor equipment conditions, analyzing data to preemptively identify signs of potential failure. The AI system was trained on the data to learn about subtle anomalies and wear patterns, enabling maintenance interventions to be timelier and need-based.

This transition from the traditional time-based maintenance approach, which saw equipment failure rates at 80%, to the AI-driven CBM method resulted in a significant reduction of failure rates to 40%. The effectiveness of this strategic shift is depicted in the figure below, demonstrating the stark contrast in frequency of equipment failures between the two maintenance methodologies. This change not only improved equipment reliability but also contributed to overall operational efficiency and reduced maintenance costs.

Figure 2.2 provides a comparative analysis of the impact of traditional time-based maintenance versus AI-enabled condition-based maintenance on key equipment operations metrics.

The unpredictable nature of equipment failures and unforeseen breakdowns account for a substantial portion of operational downtimes, leading to significant financial repercussions. CBM, an integral part of the industry 4.0 paradigm, leverages the power of the Internet of Things (IoT), edge devices, and smart machinery.

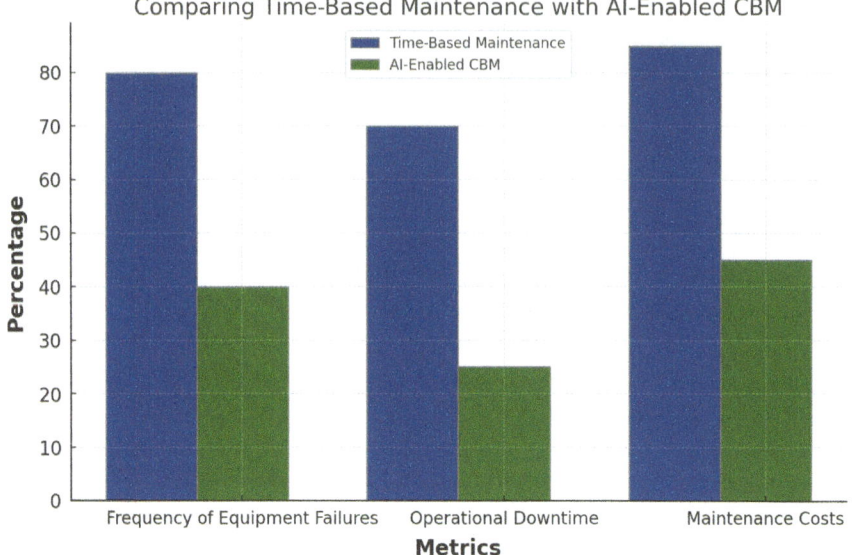

Fig. 2.2 Comparison between impact of traditional time-based maintenance and AI-enabled condition-based maintenance across key equipment operations metrics. Data is for illustrative purposes only. *Source* Created by the authors in Python Notebook

These tools continuously monitor equipment performance, and if any parameter deviates from the norm, they instantly trigger alerts.

The primary objective is to address potential issues, thereby averting costly breakdowns preemptively. AI applications enhance the efficiency of analyzing condition measurements, offering even earlier warnings of potential problems. For instance, instead of merely alerting when a machine overheats, AI can analyze patterns over time and predict when the machine is likely to overheat in the future. This predictive capability is invaluable, allowing for proactive interventions and further reducing the chances of unplanned downtimes.

The table 2.2 below provides an example of some of the use cases where AI can add value for condition-based monitoring of manufacturing equipment.

Consider the following use cases:

Vibration Analysis

Vibration analysis is a critical aspect of predictive maintenance. By monitoring the oscillatory motions of machinery, AI can detect mechanical faults like misalignment or bearing failures. This is particularly useful for assets like centrifugal pumps or motors. Over time, the AI system learns from the data, improving the accuracy of its predictions and enabling timely interventions, thereby enhancing the lifespan and performance of the machinery.

Table 2.2 Use cases of AI in condition-based maintenance for modernizing traditional maintenance practices

Use case	Description	Impact of AI applications
Vibration analysis	Monitors machinery vibrations to detect faults like misalignment or bearing failures	Enhanced prediction accuracy as the system learns from data over time
Temperature monitoring	Utilizes infrared thermography to spot temperature irregularities due to issues like misalignment	Predicts when irregularities may lead to equipment failures
Oil quality assessment	Analyzes lubricants for chemical changes or foreign particle presence	Indicates machine surface degradation crucial for assets like compressors
Sound analysis	Uses ultrasound testing to detect changes in normal operational sounds	Identifies a variety of issues, from gas leaks to under-lubrication
Electrical monitoring	Evaluates shifts in electrical parameters	Ensures the consistent operation of motors and other electrical systems

Temperature Monitoring

Temperature monitoring is another crucial component of predictive maintenance. Using infrared thermography, AI can identify temperature irregularities caused by issues like part misalignment in motors or gearboxes. These irregularities, if not addressed promptly, can lead to equipment failure. The AI system can predict when these irregularities might lead to a failure, allowing for proactive maintenance and ensuring the smooth operation of the machinery.

Oil Quality Assessment

Oil quality assessment plays a vital role in maintaining the health of machinery. AI can analyze lubricants for changes in chemical properties or the presence of foreign particles, which may indicate machine surface degradation. This is crucial for assets like compressors or vehicles, where the quality of oil can significantly impact the performance and longevity of the machinery.

Sound Analysis

Sound analysis, particularly ultrasound testing, is a powerful tool for predictive maintenance. By detecting changes in normal operating parameters, AI can identify a range of issues, from gas leaks to under-lubrication. This allows for timely interventions, helps prevent equipment failure, and contributes to the overall efficiency and safety of the manufacturing process.

Electrical Monitoring

Electrical monitoring involves evaluating shifts in electrical parameters to ensure the smooth operation of motors and other electrical systems. AI can enhance this process, providing more accurate and timely insights into potential issues. This can help prevent unplanned downtimes, improve overall operational efficiency, and contribute to the longevity of the electrical systems.

Combining CBM with AI can modernize traditional maintenance practices. It can not only reduce the frequency and impact of equipment failures but also revolutionize how industries approach asset management. As we move forward, the fusion of these technologies will undoubtedly become the gold standard, setting new benchmarks in operational efficiency and cost-effectiveness.

2.1.6.2 Summary

AI-enhanced CBM transforms maintenance from reactive to proactive, boosting productivity and reducing costs through predictive interventions.

- AI-enhanced CBM shifts the maintenance paradigm from a reactive approach, where issues are addressed after they arise, to a proactive stance, predicting and preventing potential problems before they manifest.
- By continuously monitoring equipment conditions and leveraging AI's predictive capabilities, industries can significantly reduce unplanned downtimes, ensuring smooth operations and maximizing productivity.
- AI-driven CBM allows for timely interventions based on actual equipment conditions, leading to substantial cost savings by preventing costly breakdowns and extending the lifespan of machinery.

2.1.7 Downtime Management Through AI-Powered Prediction of Failures

To maximize equipment run time utilization, the ability to predict and prevent machinery failures is becoming increasingly crucial. By adopting a predictive maintenance-based approach that applies the power of data and cutting-edge technologies to forecast potential equipment failures. This proactive strategy ensures timely interventions, minimizing downtime, and maximizing productivity.

What is Predictive Maintenance?

Predictive maintenance is a forward-thinking approach that uses data analytics, machine learning, and predictive modeling to anticipate when machinery might fail. Instead of waiting for a machine to break down or following a set maintenance schedule, this method continuously monitors equipment in real time.

Sensors gather data on various parameters, such as temperature, vibration, and pressure. Advanced AI based algorithms then analyze this data, detecting anomalies that might indicate an impending failure. By identifying these potential issues early, maintenance can be scheduled at the most opportune moment, preventing unexpected equipment downtime and enhancing overall operational efficiency.

Table 2.3 outlines various use cases and benefits of AI techniques in predictive maintenance.

Predictive maintenance has multiple benefits:

Cost Efficiency

Predictive maintenance significantly enhances cost efficiency in several ways. Firstly, by predicting failures before they occur, it allows for proactive maintenance scheduling, which can lead to substantial cost savings. Unplanned downtime can be expensive, not just in terms of repair costs, but also in lost productivity.

Furthermore, by reducing the frequency of unexpected breakdowns, predictive maintenance can also decrease the costs associated with emergency repairs, which are often more expensive than scheduled maintenance due to the need for urgent parts delivery or overtime labor. Lastly, the improved lifespan of equipment resulting from effective predictive maintenance reduces the need for costly equipment replacement, contributing to long-term cost efficiency.

Safety

Safety is a critical priority in industrial environments, and predictive maintenance plays a key role in upholding high safety standards. By forecasting potential equipment failures before they occur, it helps to avoid machinery breakdowns that could lead to workplace accidents, thus fostering a safer work environment. This proactive approach not only safeguards employees but also helps avert costly incidents that could cause damage to the facility or harm the environment.

Furthermore, predictive maintenance enables the early detection of issues that could escalate into major failures, allowing corrective actions to be taken in a planned

Table 2.3 Use cases and benefits and AI techniques in predictive maintenance

Topic	Description	Impact of applications
Cost efficiency	Predicts failures to reduce unplanned downtimes	Leads to significant cost savings and avoids lost productivity
Safety	Forecasts potential equipment failures	Enhances workplace safety by reducing the risk of accidents
Operational efficiency	Monitors equipment health in real-time	Ensures smoother workflows and increased productivity
Machine Learning (ML)	Analyzes historical and real-time data to recognize failure patterns	Predicts future equipment breakdowns
Deep Learning (DL)	Uses artificial neural networks, especially for unstructured data	Detects anomalies in machinery sounds or other intricate patterns
Natural Language Processing (NLP)	Analyzes text data, like logs or notes	Extracts insights to help predict equipment failures

and controlled manner. This proactive risk management approach reinforces a culture of safety, where potential hazards are addressed before they become serious problems, rather than reacting to emergencies after they arise.

Operational Efficiency

Predictive maintenance plays a pivotal role in enhancing operational efficiency. By providing insights into when and how equipment might fail, it allows for more effective planning of operations, ensuring smoother workflows and higher productivity. This leads to less disruption in production processes, reducing downtime and increasing output.

Furthermore, predictive maintenance improves the reliability of equipment, which means fewer interruptions and more consistent performance. Finally, the data-driven nature of predictive maintenance allows for continuous improvement, as the insights gained from the analysis can be used to optimize maintenance strategies and procedures, leading to further efficiency gains.

AI techniques for Preventative Maintenance

AI can be used for predictive maintenance using real time data. AI techniques, including Machine Learning (ML), Deep Learning (DL), and Natural Language Processing (NLP), are employed to analyze, interpret, and predict outcomes based on complex data collected from the manufacturing equipment.

Machine Learning (ML)

ML algorithms analyze both historical and real-time data from equipment sensors, learning to recognize patterns linked to equipment failures. This enables them to predict future breakdowns, allowing for timely intervention and maintenance. Furthermore, ML can adapt to changing conditions, continuously updating its models as new data is collected. This means that the predictions become more accurate over time, further enhancing the effectiveness of predictive maintenance strategies.

Deep Learning (DL)

DL is a more intricate form of ML, uses artificial neural networks to analyze data. It's especially useful for handling vast amounts of unstructured data, like audio or images, which can be used to detect anomalies in machinery sounds or visual signs of wear and tear. DL can model complex relationships in the data, making it particularly effective for tasks such as anomaly detection.

Furthermore, DL can learn to recognize subtle patterns that might be missed by other methods, leading to earlier and more accurate predictions of equipment failures. This makes DL a powerful tool for predictive maintenance, enabling more effective and timely interventions.

Natural Language Processing (NLP)

NLP analyzes text data, such as maintenance logs or operator notes, extracting insights that can help predict equipment failures. This can provide a valuable additional layer of information, complementing the data obtained from sensors. For example, an operator's note about an unusual noise or a recurring issue could be an early warning sign of a potential failure.

By incorporating this information into the predictive maintenance model, NLP can enhance the accuracy of predictions and provide a more holistic view of equipment health.

Table 2.4 details industry-specific applications of predictive maintenance using AI, showcasing how different sectors leverage AI technologies to optimize maintenance processes and outcomes.

Use Cases:

Automotive Industry

IBM has been a pioneer in leveraging AI for vehicle maintenance in the automotive industry. By partnering with automotive manufacturers, IBM uses AI to analyze data from vehicle sensors, predicting potential issues that could lead to breakdowns. This proactive approach enhances vehicle safety by identifying problems before they become serious, potentially preventing accidents.

It also improves vehicle longevity by ensuring that parts are replaced or repaired before they fail, reducing wear and tear. Furthermore, this predictive maintenance approach can lead to cost savings for both manufacturers and consumers, as it can prevent costly repairs and extend the lifespan of the vehicle.

Tech Platform

IBM's Watson IoT platform is another excellent example of AI application in predictive maintenance. The platform uses AI to analyze data from various IoT devices, predicting potential failures that could lead to downtime. By identifying issues before they cause device failure, the Watson IoT platform can reduce downtime, improving the reliability and performance of these devices.

This not only enhances user experience but also extends device lifespan, leading to cost savings. Moreover, the insights gained from this predictive maintenance can be used to inform the design and manufacturing of future devices, leading to more durable and reliable products.

Table 2.4 Industry-specific applications of predictive maintenance using AI

Industry	Use case	Example/Application
Automotive industry	Predict vehicle breakdowns	IBM partnerships: Enhances vehicle safety and longevity
Tech platforms	Predict failures in IoT devices	IBM's Watson IoT: Reduces downtime and extends device lifespan
Energy sector	Predict failures in wind turbines	Siemens Gamesa: Reduces downtime and increases energy production

Energy Sector

In the energy sector, companies like Siemens Gamesa are adopting AI for predictive maintenance of wind turbines. These turbines are equipped with numerous sensors that collect data about their operation. AI algorithms analyze this data to detect anomalies and forecast potential failures. This predictive approach reduces downtime, ensuring that the turbines are producing energy as efficiently as possible.

Furthermore, by preventing catastrophic failures, the use of AI in predictive maintenance can extend the lifespan of these turbines, leading to increased energy production over time and improved return on investment for wind energy projects.

AI-driven predictive maintenance is not just the future; it's the present. As industries continue to evolve, the integration of AI into maintenance strategies will be paramount for companies aiming to stay ahead of the curve, ensuring efficiency, safety, and cost-effectiveness.

2.1.7.1 Summary

AI-driven predictive maintenance transforms industries with proactive strategies, reducing downtimes and enhancing efficiency.

- AI-driven predictive maintenance moves industries from reactive measures to proactive strategies, drastically reducing unplanned downtimes and ensuring machinery operates at peak efficiency.
- By forecasting machinery failures, industries can not only ensure a safer working environment but also achieve significant cost savings by avoiding expensive emergency repairs and operational disruptions.
- Advanced AI techniques, including Machine Learning, Deep Learning, and Natural Language Processing, are pivotal in analyzing complex data sets, making accurate predictions, and upgrading maintenance strategies across various sectors.

2.2 Computer Vision for Defect Detection, Inspection, and Quality Control

In 2020, the Global Defect Detection Market was valued at $3.4 billion and is projected to grow to $5.1 billion by 2027, with an anticipated compound annual growth rate (CAGR) of 6.6% from 2021 to 2027 [4].

In the manufacturing industry, ensuring product quality and consistency is paramount. As the industry strives for precision and efficiency, traditional methods of visual inspection and quality control often need to catch up to the demands of modern production lines. Computer vision is being adopted for product quality assessment at a rapid pace, redefining the standards of quality control in manufacturing.

Example Use Case: Superior Inspection with AI-Driven Computer Vision Techniques in the illustrative example

At a leading manufacturing facility, a shift was made to employ AI-driven computer vision for visual defect detection and part inspection. This advanced system, powered by deep learning algorithms, automated the inspection process, allowing for rapid and highly accurate analysis of products. The AI's capability to detect even minor defects led to a remarkable increase in accuracy, from the traditional rate of 75% to an impressive 95%. Furthermore, the speed and efficiency of the inspection process were significantly enhanced, with AI achieving a 92% rate, surpassing the traditional method's 60%. Consistency, another critical factor in quality control, also saw notable improvement, increasing from 70% with traditional techniques to 90% with AI.

As the figure below illustrates, these enhancements in inspection methods through AI have not only improved the precision of defect detection but also streamlined the entire quality assurance workflow.

Figure 2.3 demonstrates the superior accuracy, speed, and consistency of AI-driven computer vision techniques compared to traditional inspection methods in manufacturing.

Computer vision empowers quality teams and machine operators on the factory floor to make informed decisions based on visual data. By replicating human vision capabilities, particularly through advanced deep learning techniques, computer

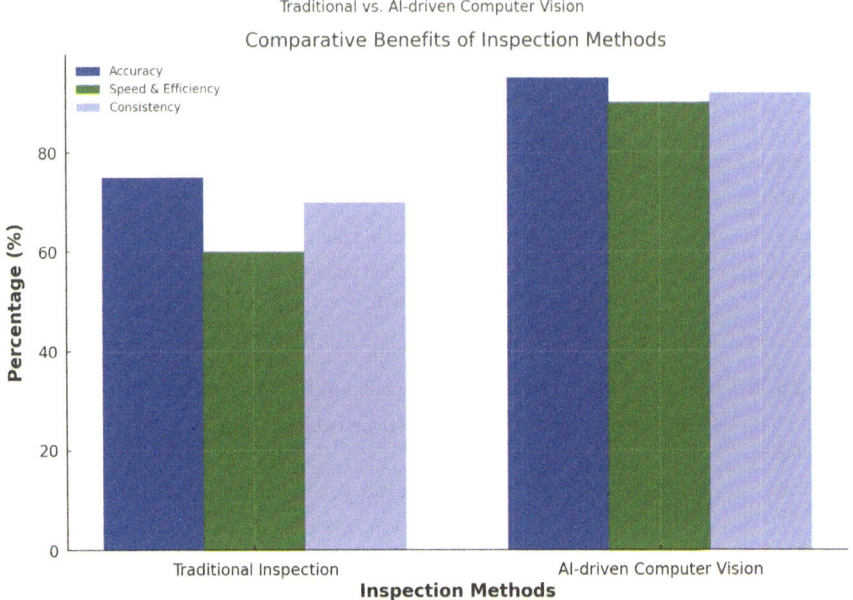

Fig. 2.3 Comparative benefits of inspection methods. Data is for illustrative purposes only. *Source* Created by the authors in Python Notebook

vision has become a vital tool for automated visual inspection and defect detection. These technologies enable systems to automatically identify anomalies, evaluate product quality, and verify that each item meets established standards, ensuring consistency and precision in the manufacturing process.

Table 2.5 provides an overview of AI-based visual inspection techniques in manufacturing.

The significance of integrating computer vision into manufacturing processes lies in its ability to enhance accuracy and speed. Traditional inspection methods, often reliant on human judgment, are prone to errors and can be time-consuming. In contrast, AI-driven visual inspection systems can process vast amounts of visual data in real time, identifying even the minutest of defects that might escape the human eye.

Table 2.6 summarizes the key benefits of integrating AI-based computer vision techniques.

One notable use case that underscores the potential of computer vision in manufacturing is its application in the automotive industry. Manufacturers are leveraging deep learning algorithms to inspect car parts, ensuring that each component, from the smallest screw to the largest panel, meets the stringent quality standards.

For instance, a computer vision system can swiftly scan a car's paint job, detecting and highlighting any imperfections, be it a tiny scratch or a slight discoloration. Such automated inspections not only ensure that the final product is flawless but also significantly reduce the time taken for quality checks.

The integration of computer vision techniques, particularly deep learning, into manufacturing processes marks a significant leap toward achieving unparalleled quality control. As industries continue to embrace AI-driven processes, computer

Table 2.5 AI based visual inspection techniques in manufacturing

Inspection technology	Description
Computer vision	Interpret visual data, mimicking human vision capabilities
Deep learning	An advanced AI technique that enhances computer vision capabilities, especially for defect detection

Table 2.6 Key benefits of integrating AI based computer vision techniques for inspection

Key advantages	Advantages of AI driven inspection over traditional approach
Enhanced accuracy	AI-driven visual inspection systems can identify minute defects that might be overlooked by human inspectors
Speed and efficiency	Processes vast amounts of visual data in real-time, optimizing both time and quality assurance
Consistency	Ensures that every product aligns with set standards, enhancing overall product quality and reliability

vision can enhance efficiency, accuracy, and consistency in product quality. The future of manufacturing is not just automated; it's visually intelligent.

2.2.1 Summary

AI-powered computer vision revolutionizes visual defect detection, enhancing accuracy, consistency, and efficiency in quality assurance.

- Computer vision, powered by deep learning, provides an innovative approach for visual defect detection, outpacing traditional human-dependent methods in terms of accuracy and consistency in product quality.
- Beyond mere defect detection, AI-driven visual inspection systems excel in the identification and classification of these defects. They can swiftly process massive volumes of visual data, pinpointing even the most subtle flaws in real time that may be missed by human eyes, optimizing both time and quality assurance processes.
- In diverse manufacturing sectors, such as electronics and semiconductors, the adoption of computer vision techniques is expanding rapidly. These AI-powered systems are setting new industry standards in quality control, ensuring that every product meets stringent criteria for excellence.

2.2.2 AI-Powered Optical Inspection to Automate Surface Defect Detection

In manufacturing, ensuring product quality is paramount. A lapse in quality can tarnish a brand's reputation, lead to compliance issues, and even result in significant financial losses. Traditionally, the responsibility of defect detection rested heavily on human inspectors. They would visually examine each product to ascertain its adherence to quality standards.

However, this manual approach, while straightforward, could be more challenging. The human eye, despite its precision, can miss minute inconsistencies or faults. Moreover, documenting and communicating these defects can be inefficient, leading to production bottlenecks, increased costs, and potential customer dissatisfaction.

Example Use Case: Enhancing Product Quality with AI-Powered Optical Inspection in the illustrative example

At a state-of-the-art manufacturing facility, an initiative was launched to implement AI-driven optical inspection for detecting surface defects. The system employed advanced AI algorithms to automate the inspection process, analyzing products with a precision and consistency far exceeding human capabilities. Trained to identify even the smallest imperfections, the AI system significantly enhanced the accuracy of defect detection. As a result, the inspection precision surged to 90%, compared to 70% with traditional human-based visual inspection, as illustrated in the figure below. This transition from conventional methods to AI-powered optical inspection not only boosted defect detection accuracy but also streamlined quality assurance, reduced the chances of defective products reaching customers, and helped protect the brand's reputation.

Figure 2.4 presents a comparative analysis that highlights the superior efficiency of AI-powered optical inspection over traditional methods.

Advanced manufacturing businesses are now applying AI to modernize their practices for defect detection. By integrating computer vision with AI, manufacturers can automate visual quality inspections, thereby reducing the costs associated with quality management. This combination not only streamlines the inspection process but also enhances its accuracy. For instance, using cameras and computer vision, defects as minute as a misprinted barcode or a tiny crack in a metal piece can be detected in real time. Such precision is often beyond human capability, especially when production rates are high.

Table 2.7 provides an in-depth look at how AI-driven optical inspection is revolutionizing quality control in manufacturing.

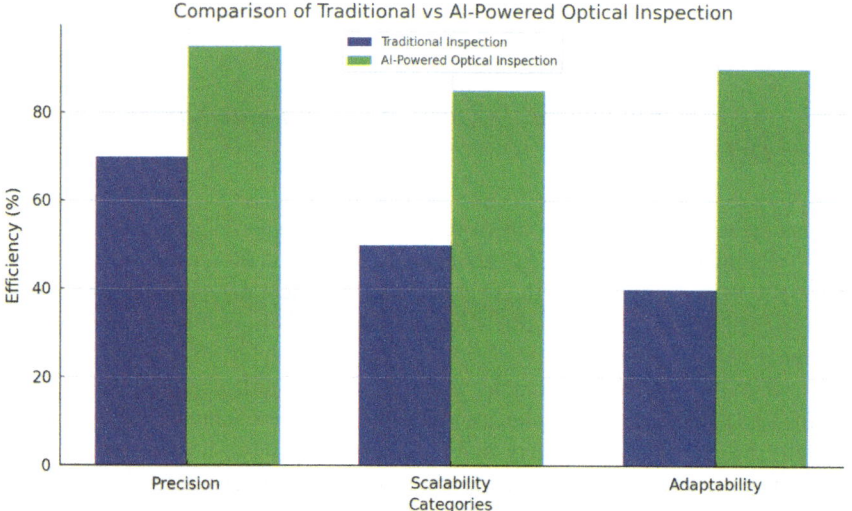

Fig. 2.4 A comparative analysis showcasing the efficiency of traditional inspection against AI-powered optical inspection in terms of precision, scalability, and adaptability. AI-driven methods consistently outperform traditional methods across all categories. Data is for illustrative purposes only. *Source* Created by the authors in Python Notebook

Table 2.7 AI-driven optical inspection: revolutionizing quality control in manufacturing

Key aspects	Description	Solution strategies/ Implementation
Traditional inspection	Relies on human inspectors for quality checks	Can miss minute defects; potential for errors and inefficiencies
AI-powered optical inspection	Integrates computer vision with AI for automated visual inspections	Swiftly detects multiple flaws; surpasses human capabilities in precision; adaptable and scalable
Impact	Streamlines the inspection process, reduces costs, and enhances accuracy	Reduced wastage, consistent product quality, optimized production efficiency, and promotes customer trust
Use cases	Packaging lines checking "Best By" dates; metal piece inspections for cracks	Ensures product excellence and prevents defective products from reaching customers

Benefits of the AI-powered approach

It eliminates the potential for human error, ensuring that fewer defective products slip through the cracks. The AI algorithms can detect multiple flaws in a product within seconds, a feat that might take humans a considerably longer time.

While human inspectors require training for new product specifications, AI models can be retrained in a shorter span, saving both time and resources. AI-driven inspection systems can be scaled across multiple production lines, offering consistent and effective defect detection.

Figure 2.5 visually presents the workflow for implementation of AI-based anomaly detection solution.

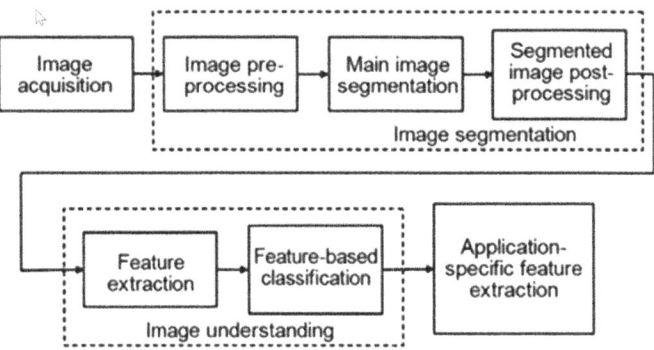

Fig. 2.5 An example workflow for implementation of AI-based anomaly detection solution. *Source* Created by the authors in Canva

Applications and use cases

In a packaging production line, computer vision can be employed to detect valid printed "Best By" dates. Any misprinted, invalid, or missing dates can be identified and rectified before the product reaches the customer. In another scenario, metal pieces moving down a production line can be inspected for cracks using specialized cameras, ensuring that only flawless pieces make it to the final product.

The fusion of AI and computer vision is setting new standards in manufacturing quality control. By automating defect detection, manufacturers can ensure product excellence, reduce wastage, and promote customer trust, all while optimizing production efficiency.

2.2.2.1 Summary

AI-driven optical inspection systems enhance defect detection, adaptability, and production efficiency, leading to significant cost savings and improved customer trust.

- AI-driven optical inspection systems surpass human capabilities in detecting minute defects, ensuring unparalleled precision in real-time, especially at high production rates.
- Unlike human inspectors who require extensive training for new product specifications, AI models can be swiftly retrained, making them adaptable to changing production requirements and scalable across multiple lines.
- By automating defect detection, manufacturers can significantly reduce wastage, ensure consistent product excellence, and optimize production processes, leading to substantial cost savings and improved customer trust.

2.2.3 Machine Vision Systems for Monitoring Product Dimensions and Specifications

A notable application of Machine Vision (MV) in manufacturing is the inspection of product dimensions. For instance, a robust MV system developed by researchers at The University of Texas Rio Grande Valley was designed to perform comparative dimensional inspection on diversely shaped samples, including those from additive manufacturing processes. The system uses blob analysis on a calibrated camera to determine various parameters of a base product, such as its perimeter, area, rectangularity, and circularity.

These parameters then serve as a standard against which other products are judged. Each subsequent product is inspected similarly and compared to the base product's parameters. A likeness score is then calculated for each product, indicating how closely it matches the base product. If the score is within a predefined threshold, the product is deemed acceptable; otherwise, it's considered defective.

Table 2.8 Application of machine vision in manufacturing to ensure precision and quality through dimensional inspection

Key aspects	Description	Solution strategies/ Implementation
Machine Vision (MV) for dimensional inspection	Uses blob analysis on a calibrated camera to determine parameters of a base product (e.g., perimeter, area)	Products are compared to the base product's parameters to ensure consistency and quality
Defect detection	Systems designed to identify minute defects in diverse product shapes	Ensures only high-quality items reach consumers; detects even millimeter-sized defects
Likeness score calculation	Score indicating how closely a product matches the base product	Provides a quantifiable measure to judge product quality; products outside the threshold are deemed defective
Integration in manufacturing workflow	MV systems integrate seamlessly into high-speed production lines	Ensures consistent product quality without slowing down manufacturing processes

Another compelling use case comes from Keyence, a leader in automation sensors. Their vision systems are designed to detect minute defects in products, ensuring that only the highest quality items make it to consumers. These systems can identify defects in a wide range of products, from cubes and cylinders to more complex shapes like sinusoidal objects. The ability to detect even millimeter-sized defects makes these systems incredibly valuable in industries where precision is crucial.

Table 2.8 shows the enhanced capabilities of MV methods across various aspects of inspection.

Manufacturing processes become more complex, and consumer expectations rise. The importance of reliable quality control systems cannot be overstated. Machine Vision can ensure consistent product quality at high-speed manufacturing production lines. MV systems that can detect finer defects, handle a broader range of products, and integrate seamlessly into the manufacturing workflow.

2.2.3.1 Summary

Machine Vision systems with ML ensure precise, high-speed quality control across diverse product lines, detecting even the smallest deviations.

- Machine Vision (MV) systems, enhanced with ML, offer unparalleled precision in inspecting product dimensions, ensuring that even the minutest deviations from the standard are detected promptly.
- Advanced MV systems can handle a diverse range of product shapes and sizes, from simple geometric forms to complex designs, ensuring consistent quality across various product lines.

- With the ability to rapidly compare products against a base standard and calculate likeness scores, ML-driven MV systems optimize the inspection process, ensuring high-speed quality control without compromising accuracy.

2.2.4 AI Based Anomaly Detection in Production Lines

Anomaly detection in manufacturing is the process of identifying data points that deviate from the expected norm. These anomalies often indicate defects in production that, if left unchecked, could result in significant losses for manufacturers. With the adoption of AI, this detection process has become more sophisticated, accurate, and timely. AI algorithms sift through vast amounts of data, identifying patterns and deviations that might be invisible to the human eye.

The primary motivation behind employing AI-driven anomaly detection is to enhance product quality and reduce wastage. Early detection of defects means fewer faulty products, leading to cost savings and increased customer satisfaction. Moreover, by identifying and addressing issues in real time, manufacturers can prevent potential machine breakdowns and ensure the safety of their staff.

The AI-based anomaly detection process begins with data collection. Sensors and cameras installed across the manufacturing facility gather data, which is then fed into machine-learning models. These models have been trained on vast datasets to recognize what constitutes "normal" and "anomaly" for specific manufacturing processes. When an anomaly is detected, alerts are generated in real-time, allowing for immediate intervention.

Table 2.9 outlines the role of AI-driven anomaly detection in manufacturing.

Example Use cases:

Foam Manufacturing

Foam manufacturing is a process that involves mixing, molding, curing, and cutting foam materials into various shapes and sizes. Foam products are widely used in various industries, such as furniture, packaging, insulation, and automotive. AI-driven image anomaly detection has been employed in foam manufacturing to ensure the consistent quality of foam products.

By analyzing images of the foam, the system can detect inconsistencies and defects that might compromise the product's integrity or performance. For example, the system can identify bubbles, cracks, holes, or uneven surfaces in the foam and alert the operators to take corrective actions. This way, AI can help reduce waste, improve efficiency, and enhance customer satisfaction in foam manufacturing.

Real-time Data Collection with Computer Vision Systems

In more complex and automated manufacturing setups, more than traditional tools like condition monitoring might be required. Here, real-time data collection with computer vision can be employed. Using machine learning models, facilities can

Table 2.9 AI-driven anomaly detection in manufacturing: enhancing quality control and operational efficiency

	Description	Implementation/Benefit
Data collection for anomaly detection	Sensors and cameras gather data throughout the manufacturing facility	Real-time data is fed into machine-learning models for immediate analysis and anomaly detection
Benefits in manufacturing	AI-driven anomaly detection techniques in foam manufacturing, real-time data collection with computer vision, and quality control	Ensures product quality, reduces wastage, identifies inconsistencies, and automates the quality control process
Anomaly detection solution architecture	Comprehensive solution involving image segmentation, image comparison, Azure integration, machine learning models, and alert mechanisms	Automates anomaly detection in foam production, integrates seamlessly with existing infrastructure, provides real-time alerts, and offers visualization tools for operational oversight
Operational efficiency and benefits	AI-driven anomaly detection systems offer real-time insights and alerts	Enables manufacturers to maintain high standards of quality, reduces staffing costs, potential yearly savings, and minimizes scrap production

leverage AI to identify patterns across the data collected from each production process, pinpointing anomalies and addressing them instantly.

For instance, computer vision can monitor the temperature, pressure, speed, and vibration of the machines and equipment, and detect any deviations from the normal range. Computer vision can also track the movement and location of the products and materials, and optimize the workflow and inventory management. By collecting and analyzing data in real time, computer vision can help improve the reliability, safety, and productivity of the manufacturing systems.

Quality Control with Computer Vision

Quality control is a crucial aspect of any manufacturing process, as it ensures that the products meet the specifications and standards of the customers and the regulators. Quality control can be challenging and costly, especially when dealing with large volumes, complex shapes, or intricate details of the products. Computer vision, combined with AI, can be used to automate the quality control process. For instance, inspecting complex parts like circuit boards manually can be time-consuming and error-prone.

Automated visual inspection using computer vision can quickly identify defects, such as missing components, misalignments, or soldering errors, ensuring that only top-quality products reach the customers. Computer vision can also measure the dimensions, colors, textures, and orientations of the products, and verify their conformity with the design specifications. By using computer vision for quality control, manufacturers can save time, money, and resources, while enhancing the accuracy, consistency, and customer satisfaction of their products.

Anomaly Detection Solution Architecture

Adastra Corporation, in collaboration with the client's Operation Technology team, developed a Proof of Concept (POC) for advanced analytics in one of the client's plants. The POC aimed to automate the detection of anomalies in foam production using image segmentation.

Figure 2.6 illustrates the detailed architecture of an AI-based anomaly detection solution along with its functional components.

Architecture and Solution:

Image Segmentation Model

The image segmentation model plays a crucial role in foam manufacturing. It identifies the cream line, a boundary that separates the liquid pour from the solid expanded foam. The position of this line is vital for consistent foam production. If the line is too high or too low, it could affect the quality of the foam. By accurately identifying the cream line, the model ensures that the foam is produced under optimal conditions, leading to a high-quality product.

Image Comparison

Images captured every 10 s are automatically compared with images from successful past runs for a specific recipe. This comparison is crucial in maintaining the quality of the foam production. A predefined tolerance level is set to ensure that the position and shape of the cream line, a critical factor in foam production, are within acceptable

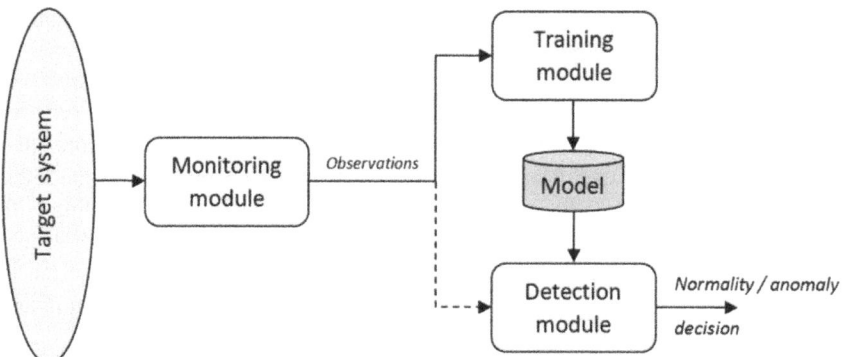

Fig. 2.6 Representative AI-based anomaly detection solution architecture. *Source* Created by the authors in Canva

limits. If any deviation is detected, it signals a potential issue that may need to be addressed to maintain the quality of the foam.

Azure Integration

New images, sent by Programmable Logic Controllers (PLCs) to a file server, are continuously monitored by Azure Logic Apps. The moment a new file is detected, the anomaly detection process is initiated. This real-time monitoring and immediate response mechanism ensure that any potential issues in the foam production process are identified and addressed promptly, thereby maintaining the quality and efficiency of the manufacturing process.

Machine Learning Model

Adastra's team developed a machine learning model using Python, which was executed via Azure Functions, to analyze the images from the foam manufacturing process. The model's algorithm was designed to determine if the position of the cream line, a critical factor in foam production, was within the client's predefined tolerance limits.

This innovative application of machine learning helps ensure the consistent quality of the foam products by enabling real-time analysis and immediate response to any deviations in the production process.

ML Algorithm

The K-means clustering algorithm, an unsupervised machine learning technique, plays a crucial role in the anomaly detection process. It extracts pixel coordinates of cream line locations in the foam production images and calculates the Euclidean distance from the average position. This calculation helps determine if the cream line's position deviates from the norm.

The model is trained using historical data and images from the SQL server, along with telemetry targets in a SQL Express database. This sophisticated use of the K-means clustering algorithm enables the system to maintain the consistent quality of foam products by identifying potential issues in real-time.

Alert Mechanism

The alert mechanism is a crucial component of the anomaly detection process. If an anomaly is detected in the foam production process, Azure Logic Apps triggers an email alert, notifying the relevant personnel about the issue. This allows for immediate action to be taken to rectify the problem. In addition to triggering alerts, both standard and anomalous images are stored in the Azure Data Lake.

The storage not only provides a record of all images for future reference but also aids in further refining the machine learning model by providing additional data for training and validation.

Tableau/Power BI Dashboard

The operational dashboard, powered by Tableau or Power BI, displays real-time data, including images and telemetry KPIs. Refreshing every 10 s, this "heads-up" display allows for recalibration and showcases trends, predicting potential errors. This dynamic visualization tool provides a comprehensive overview of the manufacturing process, enabling quick identification of anomalies and facilitating prompt corrective actions.

Operational Efficiency

Operational efficiency is significantly enhanced as production engineers, instead of continuously overseeing production lines, rely on the dashboard and anomaly alerts. They can adjust line settings during live runs if necessary, optimizing the manufacturing process. This approach not only reduces the need for constant manual supervision but also enables timely interventions, thereby improving the overall efficiency and productivity of the manufacturing operations.

Impact

The POC, typically spanning six weeks, exhibited significant potential benefits, including yearly savings of over $1 million, a 2–5% reduction in scrap, and staffing cost reduction. For a full-scale deployment from the POC, modifications and standardizations are required at the client's facilities.

The AI-driven anomaly detection in manufacturing, can be powered by various algorithms such as Convolutional Neural Networks (CNNs), Deep Learning algorithms, Xgboost, Light GBM and Support Vector Machines (SVM). By leveraging the predictive capabilities of these ML models, manufacturers can detect subtle irregularities on the production line with unparalleled accuracy. This ensures that manufacturers can maintain the highest standards of quality and efficiency, making AI an indispensable tool in the modern manufacturing landscape.

Key Takeaways:

- AI-driven anomaly detection systems offer unparalleled precision in identifying defects, ensuring consistent product quality, and reducing wastage in real time.
- By leveraging AI and machine learning models, manufacturers can receive instant alerts about anomalies, allowing for immediate corrective actions, thereby preventing potential machine breakdowns and ensuring operational efficiency.
- Automated anomaly detection can lead to significant cost savings by reducing faulty products and unplanned downtimes. Additionally, these AI systems can be scaled across multiple production lines, ensuring consistent quality control across the entire manufacturing facility.

2.3 Natural Language Processing to Extract Insights from Unstructured Text Data

NLP to Extract Insights from Manuals, Logs, Tickets, and Documentation

In the next few years, the global market for natural language processing (NLP) is expected to witness a substantial surge, growing from approximately three billion U.S. dollars in 2017 to over 43 billion by 2025 [5].

Natural Language Processing (NLP) can analyze unstructured data in manufacturing and make the communication process more efficient for humans and computers. The sheer volume of data generated in the manufacturing sector, much of which still needs to be tapped due to its unstructured nature. By leveraging NLP, manufacturers can automate processes, optimize operations, and gain a deeper understanding of customer sentiments, thereby enhancing decision-making and operational efficiency.

The NLP technology goes beyond just understanding human language; it enables computers to interpret, respond, and even predict human intentions, making it an invaluable asset in areas such as process automation, inventory management, emotional mapping, and operation optimization.

2.3.1 Example Use Case

Enhanced Efficiency by adopting NLP based bot at a manufacturing floor in the illustrative example

In a manufacturing plant, a machine operator previously faced challenges with time-consuming manual searches through extensive maintenance manuals for troubleshooting and routine maintenance, a process prone to errors. To enhance efficiency, the plant implemented an NLP (Natural Language Processing) based bot. This bot, trained on the plant's various documentation, enables the operator to quickly receive accurate, step-by-step maintenance instructions through simple natural language queries. This technological solution has significantly improved the operator's efficiency, increasing it from 60 to 90%. The benefits include substantial time savings, improved accuracy of information, reduced dependency on other teams, and optimized resource utilization. The NLP bot not only streamlines the maintenance process for the operator but also contributes to the overall operational efficiency of the plant.

Figure 2.7 visually compares the enhanced efficiency of Natural Language Processing (NLP) in processing unstructured data from various sources against traditional methods.

With tools like LLM APIs, Apple Siri or Microsoft Cortana, NLP has shown its capability in real-world applications. In the manufacturing sector, NLP can automate mundane tasks, streamline inventory management, and even map customer emotions to improve product offerings.

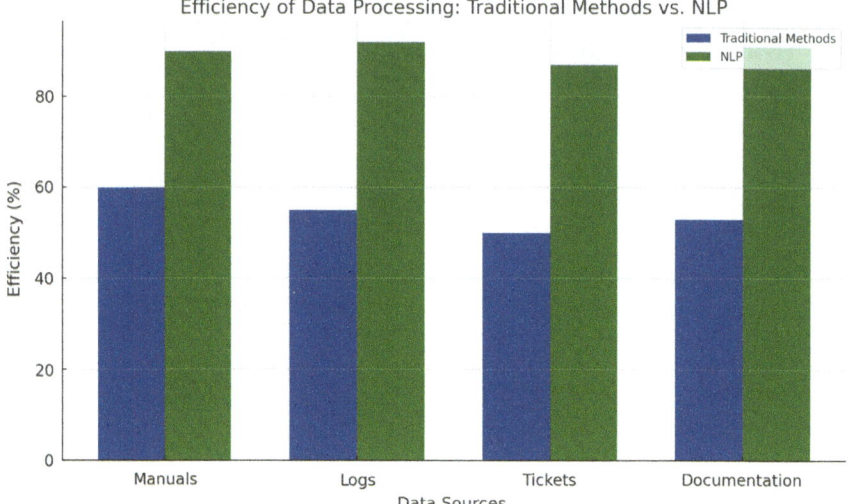

Fig. 2.7 Enhanced efficiency of NLP in processing unstructured data from various sources compared to traditional methods. Data is for illustrative purposes only. *Source* Created by the authors in Python Notebook

For instance, analysis of text data in manuals, logs, tickets, documentation and sensors in a manufacturing setup, when integrated with NLP tools, can provide real-time insights, enabling quick decision-making, improving customer experience, and preventing potential mishaps.

With an estimated 463 exabytes of data generated daily, the challenge lies in utilizing this data effectively. NLP offers a solution by processing vast amounts of unstructured data, analyzing it, and providing actionable insights. Whether it's automating processes, managing inventory, mapping customer emotions, or optimizing operations, NLP holds the key to revolutionizing the manufacturing industry. For instance, by monitoring equipment performance in real time, NLP can ensure optimal productivity and efficiency.

2.3.2   *NLP for Analysis of Maintenance Logs and Technician Notes*

In manufacturing, maintenance encompasses both preventive and corrective actions aimed at retaining or restoring the equipment's functionality. Traditionally, this maintenance data, whether preventive or corrective, is recorded in structured formats known as maintenance records, reports, or work orders.

These records are typically filled out through graphical user interfaces (GUI) and contain various fields such as record identification, asset information, maintenance activity, failure cause, and remedy. Text entry fields can be utilized for Natural Language Processing (NLP) applications as they allow operators to input unstructured text, detailing the work carried out based on their judgment.

However, the challenge arises when it comes to analyzing these unstructured text fields. Manual analysis, which involves reading each record individually, is not only tedious but also resource-intensive. This is where NLP offers a more efficient and automated approach to analyzing these text fields. By employing NLP, it becomes feasible to process and extract valuable insights from the vast amounts of unstructured text and log files in maintenance records.

Table 2.10 illustrates how leveraging Natural Language Processing (NLP) in manufacturing transforms unstructured maintenance data into actionable insights.

Consider the case study of rail infrastructure maintenance records provided by Trafikverket (Swedish Transport Administration). The data, spanning 13 years, consisted of over 10,000 records. Traditional methods would require manual reading of these records, a daunting task given the volume. However, with NLP, the analysis revealed specific patterns and insights.

Table 2.10 Leveraging NLP in manufacturing: transforming unstructured maintenance data into actionable insights

Key aspect	Description	Solution strategies/ Applications
Nature of maintenance data	Includes both preventive and corrective actions, often recorded in structured formats such as logs	Detailed records in maintenance logs and technician notes provide valuable insights into equipment functionality and necessary interventions
Challenges with unstructured text	Unstructured data entries, like text fields in maintenance records, often rely on operator judgment, making them prone to inconsistencies	Manual analysis is labor-intensive and error-prone, requiring significant resources to extract meaningful insights
NLP in action: case study	A case study analyzing 13 years of rail infrastructure maintenance data from Trafikverket, covering over 10,000 records	NLP techniques identified patterns and key insights, such as correlations between specific faults and disruptions, outperforming manual data extraction
Benefits of NLP	Streamlines data analysis, reduces human error, and improves data quality by identifying missing or critical information	Enables the detection of failure patterns, supports predictive maintenance, facilitates proactive interventions, and optimizes maintenance processes

For instance, NLP analysis identified 68 maintenance records related to cable disruptions, whereas manual extraction identified only 47. Similarly, NLP pinpointed 144 records related to computer freezes, a task impossible through manual extraction. The application of NLP to maintenance records not only streamlines the analysis process but also enhances the accuracy and comprehensiveness of the insights derived.

It eliminates the human errors associated with manual data entry and analysis. Moreover, NLP's ability to improve data quality by identifying missing or crucial information in predefined fields is invaluable.

The integration of NLP in analyzing maintenance logs and technician notes represents a significant advancement in AI-driven manufacturing processes. By automating the analysis of unstructured text data, manufacturers can gain deeper insights, optimize maintenance operations, and drive efficiency across the board. Whether it's identifying patterns in equipment failures, predicting potential malfunctions, or streamlining maintenance operations, NLP is set to revolutionize the way manufacturers approach maintenance.

2.3.2.1 Summary

Integrating IoT, AI, and NLP in predictive maintenance transforms data interaction, enhances efficiency, and minimizes downtime through automated insights and proactive interventions.

- NLP streamlines the analysis of maintenance logs and technician notes, eliminating the need for tedious manual reading and offering a faster, automated approach to extract insights from unstructured text data in AI-driven manufacturing processes.
- By employing NLP, manufacturers can significantly reduce human errors associated with manual data entry and analysis, ensuring more accurate and comprehensive insights from maintenance records.
- Integrating NLP into the manufacturing process allows for the identification of patterns in equipment failures, enabling predictive maintenance and proactive interventions, optimizing operations, and minimizing downtime.

2.3.3 AI Processes to Support Tickets and Identify Recurring Issues

AI applications such as NLP can be used in modern manufacturing processes for swiftly addressing operational challenges such as ticket support and issue identification.

Example Use Case: Reducing Equipment Downtime in Manufacturing with NLP from the illustrative example

At a manufacturing facility, an initiative to employ NLP-based automated ticketing systems was introduced. This system, equipped with advanced AI algorithms, was designed to handle tickets related to machinery malfunctions. The AI's ability to quickly interpret, categorize, and suggest resolutions for various issues transformed the ticket resolution process. The average resolution time for tickets, which traditionally took about 8 h with manual processes, was drastically reduced to just 3 h with the AI-enhanced system.

The figure below illustrates this efficiency gain, highlighting the difference in resolution times between manual and AI-driven processes across various ticket categories. This transition to AI-driven NLP ticketing systems not only expedited the problem-solving process but also improved the accuracy of issue identification and resolution. This advancement significantly reduced downtime and enhanced overall productivity.

Figure 2.8 illustrates a comparative analysis of the efficiency between manual and AI-driven processes in resolving tickets across various applications.

AI processes in ticket support involve the use of advanced algorithms and machine learning models to categorize, prioritize, and even resolve tickets automatically. These processes can analyze volumes of unstructured text data in tickets, extracting

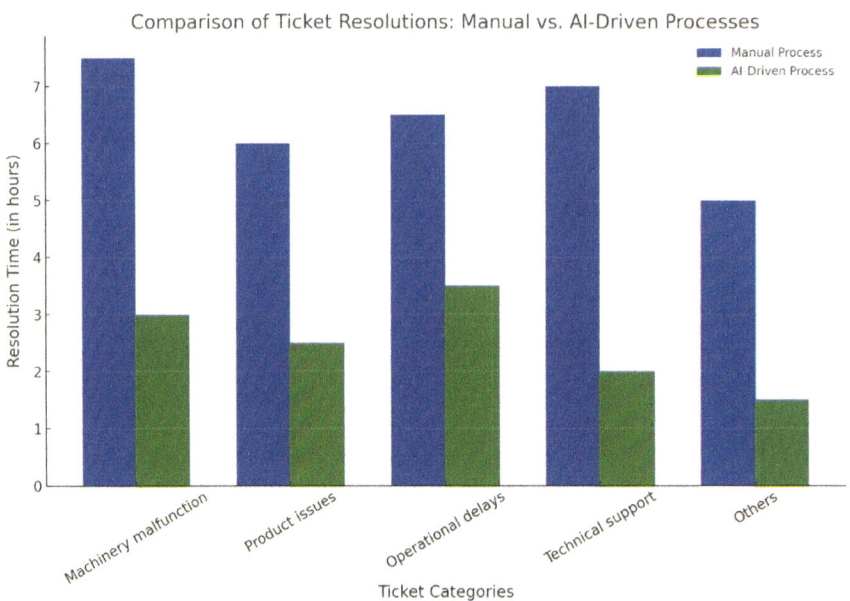

Fig. 2.8 Efficiency comparison between manual and AI-driven processes in resolving tickets across various categories. Data is for illustrative purposes only. *Source* Created by the authors in Python Notebook

relevant information and identifying patterns that might be indicative of larger, recurring issues.

The primary motivation behind integrating AI in ticket support is efficiency. Traditional ticketing systems, reliant on human intervention, can be slow and prone to errors. With the sheer volume of tickets that large manufacturing units generate, manual processing becomes a bottleneck. Moreover, identifying recurring issues manually is like finding a needle in a haystack. AI-driven processes not only speed up ticket resolution but also enhance accuracy, ensuring that critical issues don't fall through the cracks. By identifying and addressing recurring problems, manufacturers can preemptively tackle potential disruptions, ensuring smoother operations and reduced downtimes.

Table 2.11 presents how AI is used for efficient ticket support in manufacturing.

One of the primary methods is automated ticket tagging. By using Natural Language Processing (NLP), tickets are automatically categorized based on their content, ensuring they reach the right department or individual without delay. Another application is sentiment analysis on support calls. This helps in gauging customer satisfaction and identifying areas of improvement. Machine learning can be used to predict the best resolution methods for trouble tickets, reducing resolution times significantly.

Table 2.11 AI for efficient ticket support: automating processes and identifying recurring issues in manufacturing

Key aspects	Description	Solution strategies/Applications
Nature of ticketing data	Ticket data encompasses issues, requests, and feedback, often in unstructured text format	Detailed records in ticket logs provide insights into operational challenges and customer feedback
Challenges with manual ticketing	Manual ticketing systems can be slow, prone to errors, and unable to handle large volumes efficiently	Identifying recurring issues manually is challenging and can lead to unresolved critical issues
AI in action	Automated ticket tagging, sentiment analysis on support calls and predictive resolution methods	AI categorizes tickets based on content, gauges customer sentiment, and predicts best resolution methods, enhancing efficiency and accuracy
Benefits of AI	AI-driven processes speed up ticket resolution, enhance accuracy, and identify recurring problems	Enables preemptive problem-solving, reduced downtimes, and improved customer trust and satisfaction

Example Use Cases:

Automated ticket tagging in a manufacturing unit can be a game-changer. By employing AI, tickets related to machinery malfunctions can be automatically categorized based on their content. This ensures that they are instantly directed to the technical team, reducing response times and improving efficiency.

Sentiment analysis on support calls can provide valuable insights for manufacturers. By analyzing customer feedback, manufacturers can identify if a particular batch of products has consistent issues. This can guide improvements in the production process and enhance product quality.

Predictive analytics can revolutionize maintenance in manufacturing. By forecasting when a particular machine is likely to fail based on historical ticket data, preemptive maintenance can be scheduled. This not only prevents unexpected machine downtime but also optimizes the use of resources and improves overall operational efficiency.

2.3.3.1 Summary

AI enhances ticketing processes by speeding up issue resolution and enabling proactive problem management, resulting in improved customer satisfaction.

- AI-driven ticketing processes drastically reduce the time taken to categorize, prioritize, and resolve issues.
- By identifying patterns and recurring problems, AI allows manufacturers to address issues before they escalate.
- Faster ticket resolutions and the ability to preemptively tackle problems lead to improved customer trust and satisfaction.

2.3.4 NLP for Extracting Insights from Manuals and Documentation

NLP can significantly enhance the way manufacturers interact with technical manuals and documentation. As manufacturing processes evolve, so does the complexity of accompanying documentation, making quick and accurate access to information critical. NLP algorithms can automatically extract, categorize, and analyze data from various textual resources, enabling real-time insights and streamlined access to critical information. This approach allows for efficient navigation through complex documents, identifying patterns, extracting relevant data, and even answering specific queries.

By leveraging NLP, manufacturers can reduce the time spent manually searching through extensive documentation, from machinery specifications to safety protocols. The technology helps quickly locate specific data points, understand equipment details, or identify potential safety issues. This not only accelerates decision-making but also ensures that decisions are well-informed, reducing the likelihood of errors and improving overall operational efficiency.

Table 2.12 outlines the various applications of using Natural Language Processing (NLP) to extract insights from technical documentation in the manufacturing sector.

Companies like Dida and Linguamatics provide solutions that employ NLP for information extraction from technical manuals. These solutions can identify and categorize information, making it easily accessible. For instance, if a technician needs to understand a specific part of a machine, instead of manually searching through a manual, they can query the system, which then provides the relevant excerpt or even a summarized version of the information.

Example Use Cases:

Safety Protocols

In a large manufacturing plant, ensuring that all safety protocols are adhered to is paramount. NLP can quickly scan manuals to provide technicians with safety guidelines related to specific machinery. For example, if a technician needs to know how to safely operate a welding machine, they can ask the NLP system, which will then extract and summarize the relevant information from the manual.

Machinery Maintenance

When a machine malfunctions, technicians can use NLP-driven systems to find troubleshooting steps or maintenance guidelines quickly. For instance, if a machine shows an error code, the technician can input the code into the NLP system, which will then search the documentation and provide the possible causes and solutions for the error.

Regulatory Compliance

For manufacturers in regulated industries, NLP can be used to ensure that all processes align with regulatory documentation, highlighting any potential areas of

Table 2.12 Applications and benefits of NLP in extracting insights from technical documentation in manufacturing

Use case	Description	Solution strategies/Applications
Safety protocols	NLP scans manuals to provide technicians with safety guidelines for specific machinery	Ensures adherence to safety protocols, minimizing risks
Machinery maintenance	Technicians utilize NLP-driven systems to swiftly retrieve troubleshooting steps or maintenance guidelines	Speeds up machine repair and maintenance processes
Regulatory compliance	NLP checks manufacturing processes against regulatory documentation, identifying deviations	Ensures full compliance, reducing potential legal and operational risks

concern. For example, if a manufacturer needs to comply with environmental standards, they can use NLP to check their emissions data against the regulations, and identify any areas where they need to improve or report.

2.3.4.1 Summary

NLP enhances efficiency and accuracy in handling technical documentation, adapting seamlessly to evolving manufacturing processes.

- NLP drastically reduces the time taken to find and understand information in extensive technical documentation.
- By automating the extraction process, NLP minimizes human error, ensuring that the information retrieved is accurate and relevant.
- As manufacturing processes evolve and new documentation is produced, NLP systems can easily adapt, ensuring that they remain a valuable resource for operators in manufacturing floors.

2.4 Enhancing Efficiency with ARM Architecture in Smart Manufacturing Environments

The adoption of cloud computing services in the UK manufacturing industry has seen a significant increase, nearly doubling from 18% in 2018 to 32% in 2020. This surge has led to an anticipated increase in demand for the integration of ARM processor architecture in manufacturing settings. Such integration is vital for powering the computations that are crucial to contemporary manufacturing processes [6].

Utilizing ARM Processor Architecture in Manufacturing Environments

The ARM processor offers a unique blend of simplicity and power, making it ideal for a wide array of manufacturing applications. Rooted in the Reduced Instruction Set Computer (RISC) Instruction Set Architecture (ISA), ARM processors are known for their efficient use of transistors, leading to reliable performance with minimal power consumption.

This characteristic is particularly beneficial in manufacturing environments where energy efficiency and lower heat generation are crucial. Originating from Acorn RISC Machine, ARM has evolved significantly, supporting both 32-bit and 64-bit operations, and standing out with its simpler instruction set compared to more complex architectures like CISC (Complex Instruction Set Computing).

ARM ISA and Execution State

The evolution of ARM architecture took a significant leap with the introduction of Armv8, marking the transition to 64-bit processing. This change necessitated a rebranding of the original 32-bit ARM ISA to A32, while the new 64-bit ISA was named A64.

To maintain backward compatibility, Armv8 introduced two execution states: AArch32 for 32-bit operations using A32 and T32 instruction sets, and AArch64 for 64-bit operations using A64. The advancement for backward compatibility enabled ARM processors to accommodate a broader range of applications, catering to both legacy systems and cutting-edge, high-performance requirements.

ARM Architectures in Manufacturing

ARM architectures, particularly the Cortex series, have become integral to manufacturing. The Cortex series is divided into three profiles: Cortex-A for application processors, Cortex-R for high-performance real-time applications, and Cortex-M for microcontroller-based applications.

Each profile serves a distinct purpose in manufacturing: Cortex-A processors, available in both 32-bit and 64-bit versions, are well-suited for application-intensive tasks like server operations; Cortex-R excels in safety–critical applications such as medical devices and industrial control systems; and Cortex-M, designed for microcontroller, ASIC, FPGA, and SoC applications, fits perfectly into IoT applications and small, embedded systems.

Case Studies and Applications

ARM processors are used in robotics for precise control and energy-efficient operations. In network routers, ARM's low-power consumption ensures continuous, reliable performance. Furthermore, smart manufacturing devices leverage ARM's efficient processing capabilities for tasks like real-time monitoring and control. High-profile companies like Apple, Nvidia, Qualcomm, and Samsung have designed their own microarchitectures based on ARM's ISA, demonstrating its widespread adoption and adaptability in various sectors.

By understanding the evolution, execution states, and varied architectures of ARM processors, manufacturers can leverage this technology to enhance efficiency, reduce costs, and stay at the forefront of innovation in smart manufacturing environments. This integration of ARM architecture signifies a crucial step towards the future of AI-driven, smart manufacturing processes.

2.5 Digital Twin Simulation for Process Optimization

Digital Twin Simulation to Create Virtual Models of Systems and Processes for Optimization

First conceived by NASA for mirroring the condition of satellites in space using sensor data feeds, digital twins are virtual representations of physical objects or systems that leverage connectivity to continually ingest real-time data from their real counterparts. This creates living virtual models that accurately reflect their twin's state in the real world.

So, a digital twin of an aircraft turbofan engine ingests streams of sensor, maintenance and operational data during flights to simulate real-time performance. Monitoring the virtual engine helps identify anomalies, explore failure modes through simulation, evaluate hypothetical design tweaks, maximize uptime and so on—without taking physical assets offline.

In manufacturing, companies can create digital twins of key production lines, equipment like CNC machines or injection molders, products, and even entire shop floors or facilities. Feeding IoT sensor data from across the physical assets into the mirrored virtual environments enables organizations to essentially look ahead by manipulating the digital version to foresee performance, bottlenecks, disruptions and explore improvements ahead of implementation.

The manufacturing sector is expected to see considerable growth in the digital twin market by 2025, with projections expected to surpass market value of six billion U.S. dollars globally [7].

2.5.1   AI Integrated Real-Time Sensor Data into Dynamic Simulations

Creating accurate digital twin models requires setting up an extensible data infrastructure to collect and contextualize thousands of real-time data points from sensors deployed across manufacturing environments. This includes:

IoT Platform leveraging machine learning pipelines for secure connectivity, processing, and management of diverse asset data flows—

- Time series databases like InfluxDB optimized for timestamped sensor data
- Metadata standards like MQTT, OPC-UA, MTConnect for adding context
- Edge analytics and gateways enabled by AI algorithms for local aggregation and analysis.

For example, GE Aviation instruments jet engine parts with various sensors capturing temperature, vibration, fuel efficiency and other telemetry during flights. Machine learning models help make sense of the raw sensor readings by identifying correlations between related data streams across a jet engine to derive aggregate performance indicators. These are integrated and contextualized into the simulated digital twin model of the engine using MQTT connectors and a cloud-based twin platform.

AI and ML provide crucial infrastructure for reliably assimilating high velocity, heterogeneous real-time data feeds into an intelligible virtual representation in digital twins.

The reliability and business value generated from digital twins relies heavily on the integration and contextualization of massive, highly heterogeneous and noisy sensor data into meaningful simulations using artificial intelligence and machine learning techniques.

Some key focus areas where AI is pivotal—

Data Quality and Preprocessing:

Tedious data cleansing and preprocessing like handling missing values, sensor errors, temporal alignment across streams cannot scale without ML automation. Techniques like Spark/Kafka pipelines and neural network imputation improve data fidelity flowing into twins.

Anomaly Detection:

With so many data dimensions, surface-level monitoring rules miss obscure signals indicative of emerging production issues. Unsupervised ML analyzes patterns to reliably flag anomalies preceding failures captured nowhere else. This provides critical visibility.

State Estimation:

Complex interdependencies between temperature, pressure, vibration require multivariate analytics to accurately model overall equipment health and performance. AI integration modeling synthesizes various readings into reliable digital reproductions.

Simulation Calibration:

As operating conditions and sensor telemetry changes overtime, the simulations require continuous tuning for representativeness. Online machine learning dynamically calibrates digital twins improving mirroring accuracy over time using regression and time series techniques.

Inspection Analytics:

Computer vision analyzing images captured from manufacturing line camera feeds provides supplementary perspectives for virtual walkthroughs, inspection simulations and video annotations to enhance digital verisimilitude.

Achieving business value requires extensive AI infusion spanning data prep, calibration, enhancement and analytics with digital twin environments.

2.5.2 AI-Enabled Operational Changes Through Virtual Scenario Testing

Digital twins integrate multi-physics simulations of associated equipment and manufacturing processes using high-fidelity computational models unlocked through physics engines like Dassault Systèmes SIMULIA. These platforms contain libraries of prebuilt components like motors, sensors, containers etc. which can be assembled to digitally prototype production environments by specifying parameters like torque, flow rate, compression ratio and recombining parts using CAD interfaces.

Mathematical modeling grounded in scientific first principles accurately recreate complex mechanical, electrical and interdependent behaviors without needing code. Reinforcement learning algorithms can augment simulation capabilities by automatically exploring hypothetical scenarios challenging to manually enumerate. For instance, automotive assembly line simulations can ingest IoT sensor streams to train digital twin replication.

The reinforcement agent then iterates through millions of options for reconfiguring conveyor stations, optimizing mobile platform paths and evaluating paint shop robotic arm targeting configurations to maximize throughput, quality and cost metrics simulated through the digital twin. The autonomous experimentation paradigm combined with soft sensors helps surface non-intuitive optimizations unattainable otherwise.

Intel employs similar techniques combining digital twin modeling with plant data using Azure machine teaching to direct simulation iterations for optimizing semiconductor fab equipment configurations and qualifications. The intelligent exploration leads to increased yield and cycle time reductions otherwise needing months of physical equipment tuning and testing. Reinforcement learning introduces scalable, rapid scenario analysis for driving operational improvements.

AI for identification of bottlenecks and improvement opportunities

The promise of digital twins emerges from assimilating high-fidelity simulations of manufacturing processes with vast operational data streams into an integrated environment using technologies like digital threads, time series databases, and edge analytics.

The foundation of amalgamated data and computational models opens the aperture for applying artificial intelligence algorithms to uncover non-intuitive relationships and surface precision improvement opportunities through agile virtual experimentation otherwise costly and constrained.

Machine learning techniques tailored to leverage this simulation data abundance include:

1. Design of Experiments: Statistical DOE combined with Design for Six Sigma methodologies like Taguchi design systematically vary parameters and stimuli entering manufacturing process simulations to quantify operational responses using grayscale pixel embeddings. This highlights key variables and interactions driving outcomes.
2. Anomaly Detection: Isolation Forest models analyze streams of simulation telemetry for rare signals and early warning signs indicative of degrading conditions. Reviews of flagged anomalies guide proactive corrections.
3. Process Optimization: Reinforcement learning pits digital twin efficacy against iterative simulations using discounted reward functions measuring throughput, quality, changeover loss etc. to guide parameters toward Pareto-optimal operating points automatically.

4. Transfer Learning: Fine-tuning algorithms pretrained on initial simulations for analyzing niche issues like microscopic defects or minute impurities otherwise easy to overlook scales up operational learning.

By applying advanced algorithms, AI facilitates the identification and resolution of operational inefficiencies, leading to enhanced productivity. This approach not only accelerates improvements but also cultivates a culture of ongoing learning, significantly transforming the manufacturing landscape into a more agile and intelligent domain.

2.6 Key Takeaways

- AI-driven predictive maintenance shifts manufacturing from reactive to proactive maintenance, reducing unplanned downtimes by analyzing time-series data to forecast equipment failures and enabling timely interventions.
- Computer vision, powered by deep learning, automates the inspection process, significantly improving accuracy, speed, and consistency in detecting defects, part inspections, and quality control, thereby reducing reliance on error-prone manual methods.
- AI-based anomaly detection in manufacturing processes offers unparalleled precision in identifying defects and irregularities in real time, ensuring consistent product quality and operational efficiency.
- NLP enhances the analysis of unstructured text data from maintenance logs, technician notes, and documentation, providing valuable insights for predictive maintenance, troubleshooting, and optimizing operations.
- AI-driven ticketing systems improve efficiency by automating the categorization, prioritization, and resolution of issues, significantly reducing resolution times and identifying recurring problems to prevent future disruptions.
- Integrating ARM processor architecture in manufacturing environments enhances computational efficiency and energy management, supporting diverse applications from IoT devices to complex real-time control systems.
- Digital twins, virtual representations of physical systems, leverage real-time sensor data and AI to simulate, monitor, and optimize manufacturing processes, enabling predictive maintenance and operational improvements.
- AI-driven digital twin simulations facilitate scenario testing, anomaly detection, and process optimization, allowing manufacturers to identify bottlenecks and improve operational efficiency through virtual experimentation.
- Effective use of AI in manufacturing relies on integrating and contextualizing massive volumes of heterogeneous sensor data into meaningful simulations, supported by advanced AI and machine learning techniques.
- The integration of AI in manufacturing processes enables continuous learning and improvement, offering deep insights into equipment health, process optimization, and quality control, leading to enhanced productivity and cost savings.

Glossary

Anomaly Detection The process of identifying data points, events, or observations that deviate from a dataset's normal behavior.

ARM Architecture A family of reduced instruction set computing (RISC) architectures for computer processors, designed for improved energy efficiency.

RISC (Reduced Instruction Set Computer) A CPU design strategy based on simplified instructions that can be executed more quickly.

Sensor Data Information collected by devices designed to detect events or changes in an environment.

Time-Series Data A sequence of data points indexed in time order, often used in predictive maintenance.

Unstructured Data Information that doesn't have a predefined data model or isn't organized in a predefined manner.

References

1. Predictive Maintenance Market Size, Share | Industry Report, 2019–2025 [Internet]. www.grandviewresearch.com. Access date November 21, 2024. https://www.grandviewresearch.com/industry-analysis/predictive-maintenance-market
2. Fourie, O. (2024). The Interplay of IoT, Analytics, and AI in Predictive Maintenance [Internet]. Iox-connect.com. EAMS Technologies Inc. Access date November 21, 2024. https://www.iox-connect.com/journal/the-interplay-of-iot-analytics-and-ai-in-predictive-maintenance-transforming-data-interaction
3. FATFINGER. (2024). Real-world predictive maintenance: Case studies and success stories—SEE Forge creators of FAT FINGER [Internet]. SEE Forge creators of FAT FINGER. Access date November 21, 2024. https://fatfinger.io/predictive-maintenance-use-cases-triumphs-of-predi/
4. Defect detection market size, share & forecast 2021–2027 [Internet]. *KBV Research.* Access date November 21, 2024. https://www.kbvresearch.com/defect-detection-market/
5. Global natural language processing market 2017–2025 [Internet]. *Statista.* [access date 11/21/2024]; Available from: https://www.statista.com/statistics/607891/worldwide-natural-language-processing-market-revenues/
6. Manufacturing industry: use of cloud computing services 2020 [Internet]. *Statista.* Access date November 21, 2024. https://www.statista.com/statistics/475592/manufacturing-industry-uses-of-cloud-computing-services-by-entrprises-uk/
7. Global digital twin market by industry 2025 [Internet]. *Statista.* Access date November 21, 2024. https://www.statista.com/statistics/1296187/global-digital-twin-market-by-industry/

Chapter 3
AI and Advanced Analytics Applications

Abstract This chapter examines how manufacturers can leverage AI and advanced analytics to drive systemic improvements across operations. Building on foundational AI applications, it explores the transformative potential of integrating Generative AI and other advanced technologies into manufacturing processes. Key sections focus on implementing AI-enabled assistance, recommendation, and autonomous systems. Assistance systems, powered by large language models, enhance manual tasks such as programming and maintenance, while recommendation systems analyze complex datasets to optimize decision-making and operations. Autonomous systems use self-adaptive technologies for real-time process optimization, improving efficiency and reducing human error. The chapter also discusses how manufacturers can utilize big data analytics and machine learning to detect patterns, develop predictive models, and uncover cost-saving opportunities. AI-driven simulations are presented as tools to identify and resolve bottlenecks, ensuring smoother production workflows. Detailed case studies illustrate successful AI implementations, including predictive maintenance, quality control, and human resource optimization. The chapter provides step-by-step checklists for integrating AI technologies into manufacturing environments, addressing challenges such as data security, workforce readiness, and ethical considerations. By adopting these advanced technologies, manufacturers can achieve enhanced efficiency, innovation, and competitiveness, paving the way for a new era of data-driven industrial excellence.

Keywords Generative AI · Manufacturing optimization · AI Assistance systems · Recommendation systems · Autonomous systems · Process optimization

In the prior chapter, we explored various applications of AI that are transforming core manufacturing processes—from computer vision for quality control to predictive maintenance and natural language processing (NLP) for unstructured data analysis. These AI systems enable enhanced defect detection, predictive analytics, maintenance optimizations, and more.

B. Sarkar and R. K. Paul, *AI for Advanced Manufacturing and Industrial Applications*,
https://doi.org/10.1007/978-3-031-86091-1_3

Now, in this chapter, we build upon those AI foundations to explore how manufacturers can extract even greater value by leveraging AI and advanced analytics applications to enable data-driven decisions across operations.

While AI instruments and upgrades discrete manufacturing processes, the true potential is unlocking optimized, strategic decisions through manufacturing-wide data. By applying AI across siloed datasets, generating insights with advanced analytics, and leveraging recommendations from generative AI models, manufacturers can drive systemic improvements.

3.1 Adopting Generative AI and Other AI Applications in Manufacturing

The integration of Generative AI in manufacturing not only boosts operational efficiency but also drives innovation and competitive edge, paving the way for a new era of industrial advancement.

One of the key components of Generative AI is Generative Adversarial Networks (GANs). GANs are a type of AI model that uses two neural networks, a generator and a discriminator, to produce new data. The generator creates new data, while the discriminator evaluates the data for authenticity. In the context of manufacturing, GANs can be used to create realistic simulations of manufacturing processes, which can help optimize these processes and improve efficiency.

Another important component of Generative AI is Variational Autoencoders (VAEs). VAEs are a type of generative model that can learn to encode data in a lower-dimensional space and then generate new data from this space. This can be particularly useful in manufacturing for tasks such as anomaly detection, analyzing documents and manuals, and customer support, where the model can learn to recognize normal operational data, answer questions, and even identify any data that deviates from this norm.

Figure 3.1 explains the three key applications for Generative AI encompassing Assistance Systems, Recommendation Systems, and Autonomous Systems to modernize manufacturing practices, enhancing efficiency, innovation, and competitiveness.

There are three main application types:

1. **Assistance Systems**:

Assistance systems, like ChatGPT, are designed to enhance manual activities such as programming or machine maintenance. They use generative AI models to provide real-time guidance, suggest improvements, and even automate repetitive tasks. By integrating assistance systems into their operations, manufacturers can improve efficiency, reduce errors, and free up their workforce for more complex tasks.

Generative AI, particularly Large Language Models (LLMs), are increasingly being used in assistance systems. These models, such as OpenAI's GPT-3, GPT3.5 and GPT-4, GPT 4o. Open AI o1, Claude 3.5 Haiku, Mistral 7b, Falcon 180B,

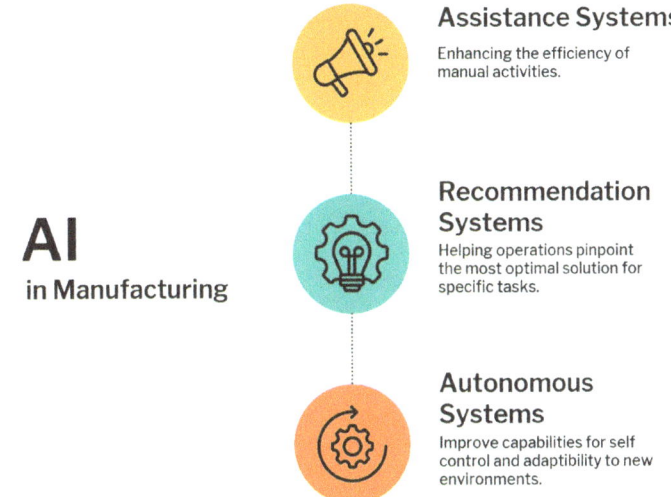

Assistance Systems

Enhancing the efficiency of manual activities.

Recommendation Systems

Helping operations pinpoint the most optimal solution for specific tasks.

Autonomous Systems

Improve capabilities for self control and adaptibility to new environments.

Fig. 3.1   AI applications in manufacturing. Created by the authors in Canva

Google's LaMDA and PaLM (the basis for Bard), Hugging Face's BLOOM and XLM-RoBERTa, Nvidia's NeMO LLM, XLNet, Cohere, and GLM-130B, are trained on vast amounts of data to mimic human behavior by performing various tasks.

2. **Recommendation Systems**:

Recommendation systems aid in identifying optimal solutions for specific tasks, such as maintenance or quality improvement. They leverage generative AI to analyze large amounts of data and provide personalized recommendations based on patterns and relationships in the data. These systems can be used to improve decision-making and optimize various aspects of manufacturing operations.

For recommendation systems, the LLMs can analyze large amounts of custom and product data and provide personalized recommendations based on patterns and relationships in the data. For instance, a novel LLM for generative recommendation GenRec can directly generate the target item to recommend, rather than calculating the ranking score for each candidate item one by one as in traditional discriminative recommendation.

3. **Autonomous Systems**:

Autonomous systems are capable of self-control and adaptability. They use generative AI to learn from their environments and adapt their behavior accordingly. These systems are used in tasks like synthetic training data generation and material handling automation. By using autonomous systems, manufacturers can automate complex tasks, improve efficiency, and reduce the risk of human error.

For instance, an innovative approach is to apply self-adaptive agents using LLMs within multi-agent systems (MASs). The methodology can be anchored on the

MAPE-K model, which is renowned for its robust support in monitoring, analyzing, planning, and executing system adaptations in response to dynamic environments.

3.1.1 How to Implement and Scale AI Applications in Manufacturing?

To fully utilize the capabilities of generative AI, manufacturers need to follow a systematic approach that includes—

Diagnosis

In the first step, manufacturers need to carry out a thorough diagnosis of their current state of operations. This involves a comprehensive analysis of production data, reviewing operational efficiency metrics, and assessing workforce capability gaps where AI could assist.

Design

After the diagnosis phase, manufacturers need to define their AI targets, strategy, and roadmap. This could involve setting specific goals for AI implementation, such as improving production efficiency or reducing waste. The strategy should outline how AI will be used to achieve these goals, and the roadmap should provide a timeline for implementation.

Engineering

With a thorough design completed, manufacturers can proceed to the engineering phase. This could involve training AI models on manufacturing data, developing algorithms for process optimization, or creating AI-powered tools for predictive maintenance. In addition, manufacturers should also implement capability-building programs to upskill their workforce and prepare them for the integration of AI technologies.

Implementation

Once the AI solutions have been developed, they need to be tested and validated. This could involve implementing the solutions in pilot areas of the production network and monitoring their performance. Any issues or challenges encountered during this phase should be addressed, and the solutions should be refined as necessary.

Scaling

After the AI solutions have been tested and validated, they can be rolled out across the entire production network. This could involve integrating the AI solutions into the existing manufacturing systems, training the workforce on how to use the new tools, and continuously monitoring and refining the solutions to ensure they are delivering the expected benefits.

Adopting AI solutions in manufacturing has multiple challenges, but it offers a path toward more efficient, innovative, and responsive manufacturing practices. In the below section the key challenges are discussed.

3.1.2 Key Challenges and Solutions for AI System Implementation

Data Security and Privacy

Ensuring data security and compliance is key when adopting AI. Manufacturers should thoroughly assess vulnerabilities in connected systems that could allow cyber-attacks to steal data or inject false inputs into AI models.

To secure AI data flows and maintain data integrity in manufacturing, implementing safeguards such as encryption, access controls, anomaly detection, and incident response plans is crucial. Vigilance against emerging threats and fostering a strong security culture allows manufacturers to utilize AI while minimizing potential vulnerabilities. Regular audits, especially with the help of external consultants, can help identify and address security gaps.

Taking a proactive approach to AI cybersecurity not only safeguards data but also ensures the reliability of AI outputs, reducing the risk of compromised results and maintaining the trustworthiness of AI-driven operations.

Skilled Workforce

While AI promises to augment human capabilities, the workforce still needs new competencies to maximize these technologies. Manufacturers should assess talent gaps in data science, machine learning engineering, analytics translation, and interfacing with AI systems on the factory floor.

Investing in robust training programs, cross-training data and operations teams, and nurturing partnerships with technical experts from academia or startups can enable manufacturers to develop, manage, and derive full value from AI tools. Cultivating talent and partnerships lays the human foundations to properly build, deploy, and act upon advanced manufacturing technologies.

Ethical and Responsible Use

Implementing AI demands proactive evaluation of potential wider impacts beyond just efficiency gains. Manufacturers should assess if AI systems could unlawfully displace workers or embed discriminatory biases compromising safety.

Mitigating risks upfront ensures responsible AI adoption. This involves efforts such as maintaining human oversight of automated decisions, enabling explainability into AI model outputs, ensuring training data is unbiased and representative, conducting environmental and community impact reviews, and engaging transparency and accountability around AI systems.

While technical fixes help, the deeper solution involves cultivating an ethical AI culture driven from the top down. Manufacturing organizations need to enact policies and oversight processes and exhibit behaviors that uphold impartiality and equitable decision-making.

3.1.3   AI-Enabled Assistance Systems in Manufacturing

AI-enabled assistance systems in manufacturing can significantly improve how factories operate, enhancing efficiency, reducing errors, and optimizing various processes. These AI-powered assistance systems in manufacturing are advanced technologies that leverage artificial intelligence to improve various aspects of manufacturing processes. These systems encompass a range of applications, from predictive maintenance to factory layout optimization. They rely on machine learning, deep learning, and other AI technologies to analyze data, automate tasks, and provide actionable insights through virtual assistance.

Figure 3.2 shows the two puzzle elements that AI-enabled Assistance Systems in Manufacturing seek to address.

3.1.3.1   AI-Driven Data Analysis in Manufacturing

These systems work by collecting and analyzing vast amounts of data from various manufacturing processes. For example, sensors can capture real-time data on machine performance, which AI models analyze to predict maintenance needs or identify potential defects. Generative AI tools, similar to ChatGPT, can also automate tasks like code generation for machine programming, thus reducing the manual workload and enhancing efficiency.

By leveraging data and algorithms to augment human capabilities, these manufacturing AI systems deliver immense value. They significantly improve efficiency and productivity, as seen in an example where an automotive supplier experienced a 21% productivity boost after deploying AI, including a scrap adviser and a pump health

Fig. 3.2 AI-enabled assistance systems in manufacturing. Created by the authors in Canva

monitor. These systems also enhance the accuracy of tasks like quality inspections and support predictive maintenance, reducing downtime and operational costs. Additionally, AI helps in optimizing factory layouts and processes, adapting to changing demands and environmental conditions.

Application in Manufacturing Process Areas

- **Facilities Design and Planning**: AI can optimize the layout of manufacturing facilities for efficiency and safety, adapting to frequent changes in processes.
- **Product Design**: AI leverages generative design techniques, helping to create optimal designs for different manufacturing requirements.
- **Predictive Maintenance**: AI analyzes data from maintenance logs to predict future machine behaviors, reducing downtime and maintenance costs.
- **Quality Control**: AI-driven visual inspection systems detect defects, improving product quality and reducing the need for manual inspections.
- **Environment, Health, and Safety (EHS)**: AI such as deep learning mines historical safety incident data to pinpoint risks. For instance, ASUS AICS developed an AI solution that improves worker safety by identifying and preventing potential risks through deep learning models.

To successfully implement AI-driven assistance systems in manufacturing, it's essential to focus on the three P's: problem, persona, and process. First, clearly identify a specific manufacturing challenge or opportunity where AI can deliver value. Then, involve the appropriate subject matter experts and key stakeholders who will play a crucial role in ensuring the project's success. Lastly, establish a well-structured process that outlines the best approach for applying AI to solve the identified problem, leveraging AI's capabilities for maximum impact.

3.1.4 AI-Powered Virtual Assistants

Traditionally, virtual assistants have been employed in internal processes to boost employee efficiency. The enhancement of these virtual agents with AI, particularly for use in manufacturing environments, brings an additional level of efficiency and functionality. These AI-empowered virtual agents, often built on sophisticated models like generative AI, are designed to interact seamlessly with human operators. They offer crucial real-time support, guidance, and problem-solving strategies, thereby enhancing the overall effectiveness of manufacturing processes.

3.1.4.1   Applications for AI-Powered Virtual Assistants

Role of AI-Powered Virtual Agents in Manufacturing

AI-powered virtual agents act as intelligent assistants. They can understand natural language inputs, process complex data, and provide relevant information or recommendations. This can range from offering troubleshooting steps for machine maintenance to guiding operators through complex manufacturing processes.

Enhancing Communication and Decision-Making

These virtual agents facilitate better communication within the manufacturing environment. They can translate complex technical data into understandable insights for operators, aiding in more informed decision-making. For instance, if a machine starts showing signs of potential failure, the virtual agent can analyze the data, compare it with historical patterns, and suggest preventive measures.

Training and Knowledge Sharing

Virtual agents are also valuable in training scenarios. They can provide interactive, on-demand training to new employees using a combination of historical data and simulated scenarios. This training can be customized to the specific needs of each employee, enhancing learning efficiency.

Real-Time Problem Solving

In the event of unexpected issues or anomalies in the manufacturing process, virtual agents can offer immediate assistance. They can quickly analyze the situation, draw from a vast knowledge base, and provide recommendations or solutions to the operator.

Continuous Improvement and Adaptation

AI-powered virtual agents continually learn from new data and experiences, adapting their responses and recommendations over time. This leads to a continuous improvement cycle in manufacturing processes, where virtual agents become more efficient and effective in their guidance.

3.1.4.2   Application in Manufacturing Process Areas

- **Quality Assurance**: A virtual agent can guide operators through quality inspection processes, ensuring adherence to standards and offering corrective actions for detected anomalies.
- **Maintenance Support**: Virtual agents can provide step-by-step guidance for routine maintenance or complex repairs, enhancing the efficiency and accuracy of maintenance tasks.

AI-powered virtual agents in manufacturing serve as intelligent, adaptive, and interactive tools that enhance the capabilities of human operators. Virtual agents play a crucial role in optimizing manufacturing processes, improving decision-making, and providing real-time support, which collectively drives efficiency, reduces downtime, and ensures consistent quality in manufacturing operations. These agents, when combined with other AI-enabled systems, signify a significant advancement in the technological evolution of manufacturing.

3.1.4.3 Case Studies on AI-Powered Virtual Assistants

Case studies showcasing how companies like Precision Global, Metromont, and Rolls-Royce are utilizing the power of AI-powered virtual assistants to enhance manufacturing efficiency and drive significant ROI [1].

Precision Global and Metromont

Companies like Precision Global and Metromont have leveraged AI to improve equipment uptime, enhance quality and throughput, and minimize scrap. These efforts not only optimized their manufacturing processes but also substantially boosted their return on investment (ROI).

Rolls-Royce

Another notable example is Rolls-Royce, which has utilized AI to drive significant improvements across various manufacturing dimensions, particularly in predictive maintenance and quality assurance.

3.1.4.4 Checklist for Step-by-Step Implementation of AI-Powered Virtual Assistants

Step 1: Diagnosis

Assess the current operational status, identify inefficiencies, and pinpoint potential areas where AI can have the most impact.

Step 2: Design

Develop a clear strategy for AI deployment, which includes setting precise objectives for AI applications, from enhancing product quality to optimizing maintenance schedules.

Step 3: Engineering

Initiate the development of AI models tailored to specific manufacturing needs, and start integrating these systems with existing IT infrastructure. Focus on both hardware compatibility and software integration.

Step 4: Implementation

Pilot the AI solutions in controlled environments to validate their effectiveness. Monitor the results closely and make necessary adjustments.

Step 5: Scaling

After successful validation, roll out the AI systems across the entire manufacturing operations. Ensure continuous training and support for all users.

Step 6: Optimization

Regularly update the AI models based on new data and feedback to enhance their accuracy and effectiveness. Also, continually assess the impact of AI on production and adjust strategies as necessary [1].

These steps should guide manufacturers through the complex process of integrating AI-powered virtual assistants into their operations, ensuring that they maximize the benefits of this advanced technology.

3.1.5 AI-Enabled Recommendation Systems in Manufacturing

AI-enabled recommendation systems play a crucial role in optimizing various aspects of manufacturing operations, improving decision-making, and enhancing overall efficiency. These AI-enabled recommendation systems are advanced tools that leverage artificial intelligence, particularly machine learning (ML) and deep learning (DL), to analyze large datasets. They provide personalized recommendations based on patterns and relationships identified within the data. These systems help in identifying optimal solutions for specific tasks such as maintenance, quality improvement, and process optimization.

These systems function by continuously ingesting new process data and adapting their models to a facility's specific needs. They evaluate ideal operating parameters and suggest configuration set-points for prevailing conditions, either in an advisory mode or by directly controlling systems in a closed loop for automatic implementation. As conditions change, the AI revises recommendations, balancing tradeoffs between competing KPI targets and refining optimization logic over time driven by new data.

3.1.5.1 Impact of AI-Enabled Recommendation Systems

The integration of AI-enabled recommendation systems in manufacturing is crucial for several reasons:

- Enhanced Efficiency and Productivity: They help in predicting machine behavior, reducing downtime, and increasing throughput, thus significantly improving overall productivity.
- Improved Quality Control: By using machine vision and deep learning algorithms, these systems can detect defects and inconsistencies in production processes that humans might overlook
- Optimized Supply Chain Management: They analyze data from various sources to optimize inventory levels and reduce lead times, helping to anticipate shifts in consumer demand and adjust manufacturing output accordingly.
- Adaptive and Flexible Manufacturing Processes: AI systems can be used to optimize manufacturing processes, making them more flexible and adaptable to current and future demands

Application in Manufacturing Process Areas:

- **Product Design**: AI in generative design creates multiple design iterations, helping in creating optimal designs for different manufacturing requirements.
- **Assembly and Material Handling**: Autonomous robots equipped with AI can perform tasks such as assembly and material handling, improving efficiency and worker safety.
- **Predictive Maintenance**: AI systems can analyze historical maintenance data to predict future machine behavior, scheduling maintenance and repairs before failures occur.
- **Quality Assurance**: AI-based systems use machine vision to inspect products and identify defects to improve product quality and reduce waste.

3.1.5.2 Case Studies for AI-Enabled Recommendation Systems in Manufacturing

Siemens and Google Collaboration: Siemens has teamed up with Google to utilize AI in enhancing shop floor productivity. This partnership focuses on applying computer vision and cloud-based analytics to improve operational efficiency and decision-making on the manufacturing floor [2].

Foxconn's Use of NVIDIA Technologies: Foxconn, along with other Taiwanese electronics giants like Pegatron and Wistron, has adopted NVIDIA Metropolis for enhancing worker safety and automating manufacturing processes. This integration showcases the use of AI to streamline complex manufacturing operations and improve overall safety and efficiency [3].

3.1.5.3   Checklist for Step-by-Step Implementation of AI-Enabled Recommendation Systems

- Step 1: Initial Assessment

  - Identify Needs: Determine which areas of manufacturing could benefit most from AI recommendations, such as production optimization, maintenance, or quality control.
  - Set Objectives: Define clear, measurable goals for the AI implementation, such as reducing machine downtime or improving product quality.

- Step 2: Data Integration

  - Data Collection: Gather relevant data from multiple sources within the manufacturing process, including machine sensors, production logs, and quality control metrics.
  - Data Preparation: Clean and preprocess the data to ensure it is ready for analysis, removing any irrelevant or corrupt data.

- Step 3: Model Development

  - Choose Models: Select appropriate machine learning models based on the specific recommendation tasks at hand, whether they involve predictive maintenance, quality optimization, or production planning.
  - Model Training: Train the models using historical data, continually validating and tuning them to improve accuracy and reliability.

- Step 4: System Integration

  - Integration with Existing Systems: Ensure the AI system is compatible with existing IT and operational technology infrastructure for seamless data flow and communication.
  - Pilot Testing: Test the AI system in a controlled section of the manufacturing process to evaluate its effectiveness and make necessary adjustments.

- Step 5: Full-Scale Deployment

  - Rollout: Deploy the system across the manufacturing operations once it has been optimized in the pilot phase.
  - Monitoring and Support: Continuously monitor the system's performance and provide ongoing support and troubleshooting to address any issues that arise.

- Step 6: Continuous Improvement

  - Feedback Mechanism: Implement a feedback loop to collect insights on the AI system's performance and impact on manufacturing outcomes.
  - Iterative Optimization: Use feedback to refine AI models and expand their application to additional areas of the manufacturing process to enhance the overall efficiency and output.

3.1.6   AI-Enabled Autonomous Systems in Manufacturing

AI-enabled autonomous systems can significantly improve industrial processes, offering self-control, adaptability, and learning capabilities that significantly enhance operational efficiency and reduce human error in manufacturing.

3.1.6.1   What Are AI-Enabled Autonomous Systems in Manufacturing?

AI-enabled autonomous systems in manufacturing are self-governing systems that use generative AI to learn from their environment and dynamically adapt their behavior. These systems are employed in various tasks, such as synthetic training data generation, material handling automation, and process optimization.

3.1.6.2   How Do These Systems Work?

Autonomous systems in manufacturing utilize AI to analyze onboard sensor data, allowing them to adapt to changing conditions and optimize processes in real-time. For example, AI can play a role in optimizing factory floor layout, spotting potential safety issues, and improving process-flow efficiency. AI applications aren't limited to fabrication but extend to planning and layout optimization of facilities.

One innovative approach involves using self-adaptive agents within multi-agent systems (MASs) based on the MAPE-K model (Monitoring, Analyzing, Planning, and Executing system adaptations). This method allows autonomous systems to be responsive and adaptive in dynamic environments.

3.1.6.3   Impact of AI-Enabled Autonomous Systems in Manufacturing Applications

Autonomous systems are crucial in manufacturing for several reasons:

- Optimization and Efficiency: These systems optimize manufacturing processes, significantly reducing waste, cost, and time. Engineers can use these tools to adjust to their environments better and meet operational objectives more effectively.
- Mass Customization and Adaptability: Autonomous AI systems in a manufacturing setting allow for mass customization, delivering highly personalized products at scale. It enables the automation of nuanced management tasks and assembly lines, making manufacturing more adaptable to varying demands.

3.1.6.4   Application in Manufacturing Process Areas

- Predictive Maintenance and Process Refinement: Autonomous AI systems provide operational feedback for preventive maintenance and process refinement, such as in additive manufacturing.
- Quality Control: Utilizing sensor data, AI systems can detect defects or inconsistencies in products, improving quality control.

Solutions Strategies Implementing AI-Based Autonomous Systems

When integrating autonomous systems into manufacturing, it's important to consider:

- Low-code AI for Optimized Control Systems: Using intuitive templates and simulation environments can lower the barrier to entry for leveraging AI in manufacturing.
- Human-Trained AI: Veteran engineers can impart their expertise directly to the AI agent, allowing for a more nuanced and informed approach to control systems.
- Understanding and learning from AI: It's crucial for manufacturers to have total control and understanding of how the AI operates in their environment, ensuring transparency and adaptability.

AI-enabled autonomous systems are improving industrial processes, enabling smarter, more efficient, and adaptive manufacturing environments. As technology continues to evolve, these systems are set to become even more integral to manufacturing operations, driving innovation and productivity to new heights.

3.1.6.5   Case Studies for AI-Enabled Autonomous Systems in Manufacturing

The 2 use cases discussed below demonstrate how Foxconn and Delta Electronics revolutionized their manufacturing processes through the strategic implementation of AI-enabled autonomous systems [3].

Foxconn's Adoption of NVIDIA Technologies: Foxconn has implemented NVIDIA's AI and Omniverse technologies to supercharge its industrial digitalization. This includes the use of AI to build, simulate, and operate factory digital twins, enhancing productivity and operational efficiency across its manufacturing plants [3].

Delta Electronics: Leveraging NVIDIA Omniverse and Metropolis, Delta Electronics has successfully integrated advanced AI technologies to digitally simulate and optimize manufacturing processes, leading to improved productivity and more agile operations [3].

3.1.6.6 Checklist for Step-by-Step Implementation of AI-Enabled Autonomous Systems

Step 1: Define Objectives

Identify Opportunities: Pinpoint areas in manufacturing where autonomy can significantly improve efficiency, such as material handling, product assembly, or quality inspection.
Set Goals: Establish clear, measurable objectives for the deployment of autonomous systems, such as reducing human error, increasing production speed, or improving safety.

Step 2: Data Acquisition and Analysis

Sensor Integration: Equip machines and operational areas with sensors to collect real-time data essential for training and operating autonomous systems.
Data Analysis: Analyze collected data to understand patterns and requirements, which will guide the development of the AI models.

Step 3: System Development

AI Modeling: Develop and train AI models tailored to specific manufacturing tasks, utilizing technologies like machine learning, deep learning, and computer vision.
Simulation Testing: Use digital twins and simulation platforms, such as NVIDIA Omniverse, to test AI behaviors in virtual environments before physical deployment.

Step 4: Pilot Implementation

Small-Scale Deployment: Start with a pilot project in a controlled part of the production line to evaluate the performance of autonomous systems.
Iterative Feedback: Collect feedback and adjust AI algorithms based on their performance in the pilot phase to ensure optimal integration.

Step 5: Full Deployment

Scale Up: Gradually expand the deployment of autonomous systems across the manufacturing operations based on success metrics from the pilot.
Integration: Ensure full integration with existing manufacturing execution systems and IoT platforms for seamless operations.

Step 6: Continuous Optimization and Scaling

Monitoring and Adjustments: Continuously monitor the systems to optimize performance and adapt to new manufacturing conditions or requirements.
Expansion: Look for opportunities to leverage AI-driven autonomy in other areas of manufacturing to further enhance operational efficiency.

3.2   Applying Big Data Analytics and Machine Learning

Manufacturers are using machine learning to detect hidden patterns and trends in large production datasets. For instance, algorithms can identify subtle vibrations that indicate potential equipment failures well before breakdowns occur. Similarly, by analyzing historical defect data, machine learning can detect slight anomalies during inspections that might signal future product issues.

This capability to uncover patterns and correlations, combined with predictive modeling and refined data segmentation, lays the groundwork for AI-powered forecasting. It enables manufacturers to take proactive measures and make smarter decisions. The following section explores how machine learning can help identify subtle correlations, segment data effectively, and develop predictive models that transform small signals into actionable insights, leading to improved manufacturing operations.

3.2.1   Identifying Trends, Patterns, and Correlations with ML

AI can improve the efficiency and quality of manufacturing processes by discerning underlying trends, patterns, and correlations from vast datasets. This capability not only aids in optimizing current processes but also in predicting future outcomes, ensuring a more streamlined and efficient production line.

Machine Learning excels at analyzing large datasets to identify hidden patterns, trends, and correlations. By training on historical data, ML algorithms can recognize intricate patterns that might be too complex or subtle for human analysts to detect. This pattern recognition capability is especially valuable in manufacturing, where even minor inefficiencies can lead to significant operational challenges.

The manufacturing sector generates enormous amounts of data daily, from equipment performance metrics to quality control measurements. Identifying patterns within this data can lead to insights that drive process improvements, cost savings, and enhanced product quality. For instance, recognizing a recurring defect pattern in a product line can lead to the identification of a faulty machine or process, allowing for timely intervention. Similarly, spotting trends in equipment performance can preemptively signal the need for maintenance, reducing unplanned downtimes.

3.2.1.1   Application in Manufacturing Process Areas

Production Data Analysis

Deep learning applications in production data analysis involve sophisticated pattern recognition techniques. By integrating data from a multitude of sensors and pieces of equipment, these advanced algorithms have the capability to identify irregularities that may not be immediately apparent to the human eye. This technology is instrumental in forecasting potential equipment failures and enhancing overall production efficiency. Its predictive capabilities allow for proactive maintenance and operational adjustments, significantly reducing downtime and improving productivity.

Quality Control

The utilization of deep learning in quality control is a significant advancement in the manufacturing sector. This approach involves the analysis of high-resolution images of products using deep learning algorithms. These algorithms are trained to recognize and pinpoint various types of defects, ranging from minor aesthetic flaws to critical functional failures. By automating the quality control process, manufacturers ensure that only products meeting the highest quality standards are distributed to consumers. This not only enhances customer satisfaction but also reduces the cost associated with returns and repairs.

3.2.1.2 Use Case Example

Optimizing Quality Control with ML and Deep Learning

Deep learning, an advanced form of ML, has shown significant promise in pattern recognition within production data analysis. These algorithms can process vast amounts of data, identifying complex patterns and relationships.

At a cutting-edge manufacturing facility, an initiative was launched to incorporate ML and Deep Learning in quality control. These AI technologies were applied to analyze production data and identify underlying trends and anomalies that could affect product quality. The traditional methods of quality control had an accuracy rate of 80%.

However, with the integration of ML, this accuracy improved to 90%. Moreover, the adoption of more advanced Deep Learning algorithms further enhanced this accuracy to 97%, as indicated in the Fig. 3.3. The transition to AI-driven quality control methods not only increased the precision of defect detection but also allowed for more proactive and predictive quality assurance.

3.2.2 Segmenting Data and Developing Predictive Models

The sheer volume and complexity of data generated from manufacturing can be hard to analyze using traditional analytics. However, when properly segmented and analyzed, this data holds the key to unlocking unprecedented levels of efficiency, quality, and innovation. The application of predictive modeling can enable data decision-making from the data created from the manufacturing processes and help develop future strategies.

Predictive modeling involves using statistical algorithms and machine learning techniques to identify the likelihood of future outcomes based on historical data. It's about understanding the relationships between different data points and using them

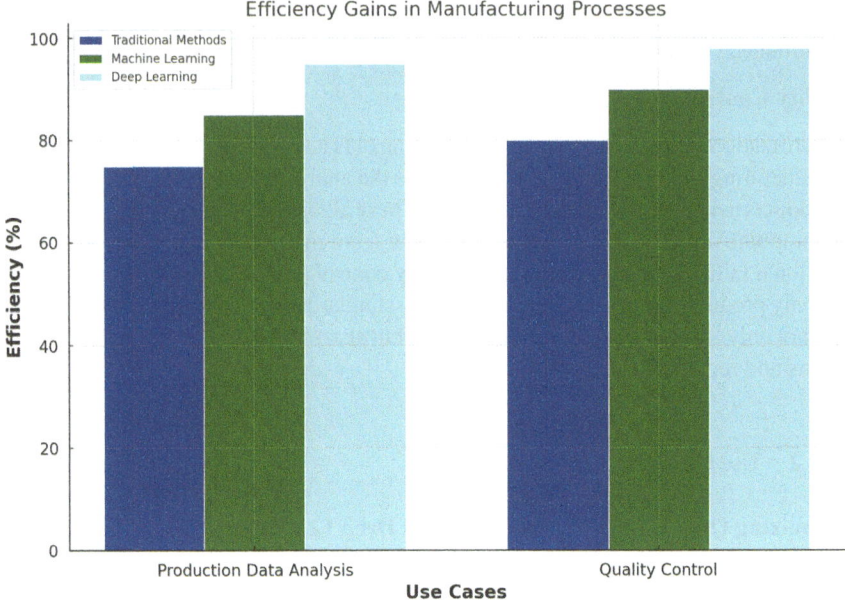

Fig. 3.3 Comparison of efficiency gains in key manufacturing processes: traditional methods versus AI-based techniques such as machine learning and deep learning. Data is for illustrative purposes only. Created by the authors in Python Notebook

to predict future trends or behaviors. Data segmentation, on the other hand, is the process of dividing a large dataset into smaller, more manageable chunks based on specific criteria, making it easier to analyze and draw insights.

Manufacturing processes are dynamic, with numerous variables at play. Predictive models can forecast potential disruptions, equipment failures, or inefficiencies, allowing manufacturers to take proactive measures. Segmenting data ensures that these models are fed relevant, high-quality data, making their predictions more accurate and actionable. In essence, it's about turning raw data into actionable insights that can drive operational excellence.

Techniques like Principal Component Analysis (PCA) and Partial Least Squares (PLS) are often used in predictive modeling. While PCA focuses on reducing the dimensionality of the data, ensuring only the most relevant variables are considered, PLS is more about understanding the relationships between observed variables and predicting outcomes.

3.2.2.1   Application in Manufacturing Process Areas

Equipment Maintenance

By scrutinizing historical data on machine performance, these advanced algorithms are adept at forecasting impending failures or maintenance needs. This predictive capability is crucial in enabling timely maintenance interventions, effectively reducing the incidence of unexpected breakdowns. The strategic implementation of these analytics leads to enhanced equipment reliability, minimized downtime, and potentially significant cost savings in long-term maintenance.

Quality Control

Utilizing data from past production cycles, AI algorithms are capable of assessing the probability of defects in current production batches. This predictive insight allows manufacturers to preemptively address potential quality issues, ensuring that the final products meet stringent quality standards. This proactive strategy not only safeguards product integrity but also significantly reduces waste and enhances customer satisfaction.

3.3   Identifying Process Optimization and Cost-Saving Opportunities

A 2017 study published by Bergur Thormundsson on March 17, 2022, sheds light on the tangible benefits of implementing AI in organizations globally. The data reveals that firms with dedicated AI leaders tend to reap more substantial rewards from AI adoption compared to those without such leadership. Notably, 19% of these organizations with an AI lead reported experiencing enhanced operational efficiency as a key benefit. This statistic demonstrates the value of strategic AI leadership in organizations, highlighting how focused guidance in AI implementation can lead to significant improvements in operational aspects [4].

With the shift to Industry 4.0, manufacturing is increasingly embracing AI and advanced analytics to navigate the complexities of modern operations. These technologies utilize sophisticated algorithms and data analysis techniques to derive actionable insights from large datasets. In manufacturing, AI and analytics play a crucial role in detecting inefficiencies, predicting equipment failures, optimizing supply chain, and enhancing overall productivity.

Manufacturing operations encompass various processes, from managing raw material quality and monitoring machine performance to responding to market demand. Even small inefficiencies can accumulate into significant costs when scaled across the entire operation. By employing AI and advanced analytics, manufacturers can gain deeper insights into their processes, identify bottlenecks, and implement data-driven solutions that deliver cost savings and operational improvements.

3.3.1   How AI Achieve Process Optimization and Cost-Saving Opportunities

AI algorithms, trained on historical and real-time data, can predict future trends, identify anomalies, and suggest optimizations. Advanced analytics, on the other hand, can run through vast datasets, segmenting and analyzing them to provide actionable insights. Together, they form a potent combination that can drive significant improvements in manufacturing processes. These technologies bring a new level of intelligence to manufacturing operations, enabling businesses to anticipate challenges and streamline their processes for maximum efficiency (Fig. 3.4).

3.3.1.1   ML Analysis of Equipment Effectiveness

In manufacturing, the efficiency and effectiveness of equipment play a pivotal role in determining the overall productivity and profitability of operations. As manufacturing processes get integrated with AI/ML can be applied to analyze and enhance equipment effectiveness.

Overall Equipment Effectiveness (OEE) is a standard metric used in manufacturing to gauge the efficiency of a machine or a production line. It takes into account three primary factors: availability (uptime vs. downtime), performance (speed of operation), and quality (number of defects). ML, with its data-driven algorithms, can analyze vast amounts of data related to these factors, offering insights and predictions that can significantly improve OEE.

Application in Manufacturing Process Areas

Predictive Maintenance:

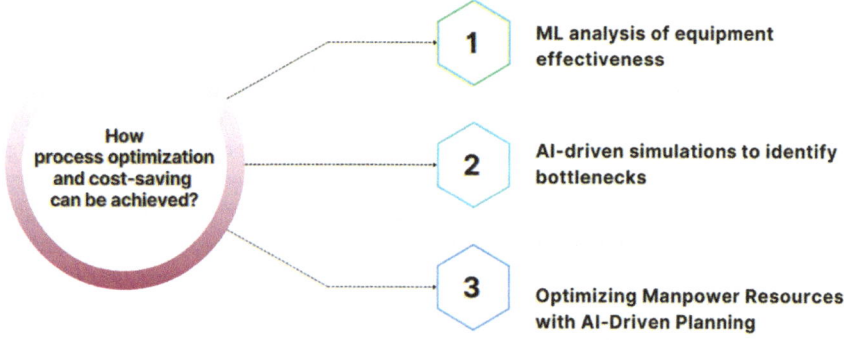

Fig. 3.4 How can process optimization and cost-saving be achieved? Created by the authors in Canva

ML algorithms analyze data from equipment to forecast potential failures, facilitating timely maintenance interventions. This predictive approach significantly reduces unplanned downtimes, enhancing operational efficiency and reliability. Manufacturers can thus ensure continuous production flow and avoid costly breakdowns, making predictive maintenance an indispensable tool in modern manufacturing settings.

Quality Control:

ML algorithms, by analyzing data from various sensors, can identify even minor deviations from quality standards during production. This level of precision ensures that every manufactured product adheres to the highest quality criteria. Consequently, manufacturers can maintain consistent product quality, reduce waste, and uphold their reputation for quality in the competitive market.

Sustainability:

By examining energy consumption and usage patterns in manufacturing equipment, ML algorithms can recommend more energy-efficient machine operation strategies. This approach not only reduces the energy footprint of manufacturing processes but also leads to considerable cost savings. Such energy optimization is crucial for environmentally conscious manufacturing and contributes to the sustainability goals of the industry.

Use Case Example

Improving Overall Equipment Effectiveness (OEE) with Machine Learning in Manufacturing

At a progressive manufacturing facility, an initiative was deployed to apply ML algorithms to boost OEE. These ML algorithms were specifically designed to analyze equipment performance data, identifying patterns and predicting potential issues that could affect availability and efficiency. Prior to the integration of ML, the facility's equipment availability rate was at 78%.

However, with the introduction of ML for equipment monitoring and predictive maintenance, this rate significantly increased to 90%, as depicted in Fig. 3.5. The adoption of ML in monitoring and analyzing equipment performance led to a notable improvement in OEE. The ML algorithms provided more accurate and timely insights into equipment health, allowing for proactive maintenance and reducing unexpected downtime.

Manufacturing industries operate on thin margins, and any downtime or inefficiency can lead to substantial financial implications. Traditional methods of monitoring equipment, often manual and reactive, may need to capture the nuances and patterns that can indicate potential issues. ML provides—

- a proactive approach.

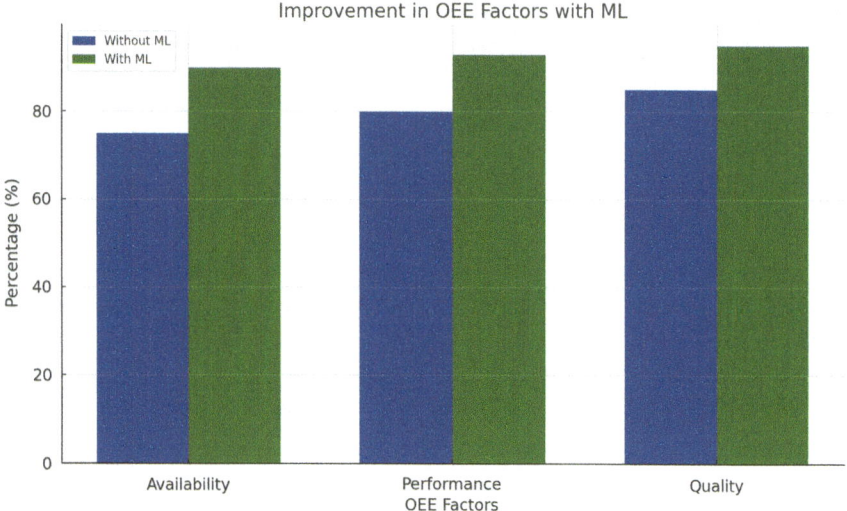

Fig. 3.5 Comparison of OEE factors' efficiency with and without the implementation of ML in manufacturing. Data is for illustrative purposes only. Created by the authors in Python Notebook

- allows manufacturers to predict and prevent problems before they occur.
- ensures that equipment operates at its optimal capacity.

By training ML models on historical equipment data, manufacturers can predict potential downtimes, identify patterns leading to inefficiencies, and even forecast maintenance requirements. These models continuously learn from new data, refining their predictions and ensuring that the insights remain relevant and actionable.

3.3.2 AI-Driven Simulations to Identify Bottlenecks

Bottlenecks caused by points of congestion during production slow down the entire process and significantly hamper productivity and efficiency. Identifying and addressing these bottlenecks is crucial for optimizing production flow and maximizing output. With the application of AI, manufacturers now have a powerful tool at their disposal: AI-driven simulations. These simulations can model complex production environments, predict potential bottlenecks, and suggest solutions to alleviate them.

AI-driven simulations are virtual models that replicate real-world manufacturing processes. By feeding these models with historical and real-time data, AI algorithms can simulate various scenarios, identify points of congestion, and predict how changes in one part of the production line might affect the entire system.

3.3.2.1   Application in Manufacturing Process Areas

Process Optimization:

Lean manufacturing simulations offer a powerful tool for identifying and addressing inefficiencies in production processes. These simulations allow manufacturers to experiment with various lean strategies in a virtual environment, helping them to pinpoint the most effective methods for minimizing waste and enhancing process flow. By adopting these simulations, manufacturers can make informed decisions about process improvements, leading to increased efficiency and reduced operational costs.

Production Planning:

AI-driven simulations are especially beneficial in complex production settings, such as those in the aerospace industry. These simulations provide a detailed modeling of intricate manufacturing environments, enabling the identification of bottlenecks that may not be easily discernible through conventional analytical methods. The advanced production planning approach helps manufacturers optimize production flows and improve overall efficiency in challenging and sophisticated manufacturing setups.

Facilities Design:

AI simulations can be used for fine-tuning production parameters, such as machine speeds and staffing levels. These simulations allow manufacturers to assess the impact of varying these parameters on production bottlenecks. By doing so, they can determine the most efficient configurations, leading to optimized production efficiency. This approach is crucial for manufacturers seeking to balance various operational aspects to achieve the highest level of productivity.

3.3.2.2   Use Case Example

Streamlining Production with AI-Driven Simulations for Bottleneck Identification

At a leading manufacturing facility, the operations team implemented an initiative to utilize AI-driven simulations to identify and address production bottlenecks. These AI-powered simulations could model the entire production environment in detail. They were capable of analyzing various production scenarios, predicting potential bottlenecks, and suggesting optimal solutions to alleviate them.

With the implementation of these AI-driven simulations, the efficiency in identifying and resolving bottlenecks improved dramatically to 95%, as shown in Fig. 3.6. This improvement was a result of the AI's ability to quickly and accurately model complex production processes quickly and accurately and identify points of congestion that were not easily detectable through traditional methods. The adoption of

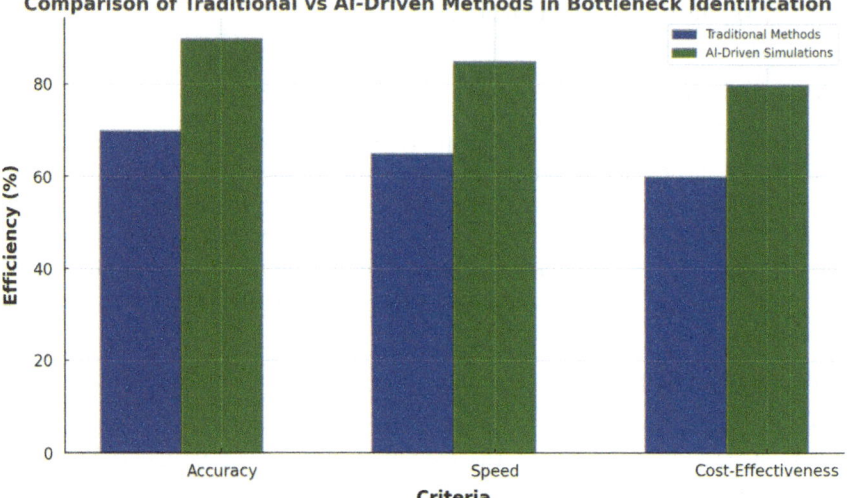

Fig. 3.6 Comparison of efficiency of traditional methods versus AI-driven simulations in bottleneck identification across different criteria. Created by the authors in Python Notebook

AI-driven simulations led to a more streamlined production flow, increased output, and significantly enhanced overall operational efficiency.

The bottlenecks, if not addressed, can lead to increased production times, higher costs, and reduced output. Traditional methods of identifying bottlenecks, often based on observation and manual analysis, might not capture the complexities of modern manufacturing environments. AI-driven simulations offer a more holistic view, allowing manufacturers to visualize the entire production process, identify weak points, and test potential solutions without disrupting actual operations.

AI algorithms create a virtual replica of the manufacturing environment using data from sensors, production logs, and other sources. This replica can then be used to run various scenarios, adjusting parameters to see how they impact production flow. By analyzing the results of these simulations, manufacturers can pinpoint bottlenecks and devise strategies to address them.

Table 3.1 provides an overview of how AI-driven simulations are utilized to identify and address bottlenecks in manufacturing. It covers key points like simulation definition, data sources, and applications, focusing on how these simulations aid in bottleneck identification, optimizing lean manufacturing practices, managing complex environments, and continuously updating parameters to improve production efficiency.

Table 3.1 Overview of AI-driven simulations for identifying and addressing bottlenecks in manufacturing

Application	Description	Impact
Simulation	Creates virtual models that replicate real-world manufacturing processes using AI algorithms	Enables visualization and prediction of outcomes without disrupting actual operations
Data integration	Utilizes data from sensors, production logs, and other real-time sources	Ensures the simulations accurately reflect real-world manufacturing conditions
Bottleneck identification	Uses AI to simulate various scenarios and identify areas of process congestion	Provides a comprehensive view of the production process, allowing targeted solutions for weak points
Lean manufacturing	Tests different lean strategies through simulation to optimize processes	Helps determine effective methods for reducing waste and improving production efficiency
Complex environment modeling	Models intricate manufacturing setups, such as aerospace or multi-step production lines	Identifies hidden bottlenecks and optimizes workflows in complex production environments
Parameter optimization	Adjusts settings like machine speeds or staffing levels to find the most efficient configuration	Allows manufacturers to achieve maximum efficiency by fine-tuning operational parameters
Continuous updates	Refines simulations based on new data and changing conditions	Ensures simulations stay relevant and accurate in predicting issues as manufacturing evolves

3.3.3 Optimizing Human Resources with AI-Driven Planning

In complex manufacturing environments, human resources remain one of the most dynamic and challenging elements to manage, even with meticulous planning for components and processes. The availability of the right number of skilled workers at the right time is crucial to maintaining a manufacturing unit's efficiency and maximizing output. With the advent of AI, manufacturers are leveraging its capabilities to optimize workforce management through intelligent resource planning.

AI-driven resource planning utilizes machine learning algorithms and predictive analytics to anticipate staffing needs based on factors such as production schedules, order volumes, and historical trends. This approach ensures that the appropriate number of workers with the necessary skills are available precisely when needed, minimizing the risks of overstaffing or understaffing. By aligning workforce allocation with real-time demands, AI enhances operational efficiency and reduces labor costs.

AI can help optimize human resources using the following techniques—

Demand-Driven Resourcing:

AI systems analyze historical order volumes and production data to forecast future resource demand. This predictive insight allows manufacturers to proactively adjust their resource levels, ensuring they are well-equipped with the necessary workforce during peak order periods. The strategic staffing approach helps in maintaining operational efficiency and meeting production demands without overstaffing or understaffing.

Skill-Based Allocation:

AI application in the skill-based allocation of the workforce can be very impactful. AI algorithms can evaluate the skills and competencies of the workforce, assigning them to roles where their expertise is most beneficial. This intelligent allocation ensures that specialized tasks are undertaken by appropriately skilled workers, enhancing the quality and efficiency of work while optimizing workforce utilization.

3.3.3.1   Use Case Example

Enhancing Operational Efficiency with AI-Driven Resource Optimization

On a manufacturing workshop floor, an initiative to adopt AI planning for resource optimization was set in motion. Leveraging sophisticated AI algorithms, the approach involved a thorough analysis of production requirements, workforce capabilities, and various operational metrics. The AI system was adept at predicting precise resource needs and aligning the right number of skilled workers with specific production schedules.

The shift to AI-driven resources led to a notable increase in efficiency, reaching 85% in terms of cost savings, as depicted in Fig. 3.7. This marked improvement was attributed to the AI's precision in forecasting resource requirements, which in turn minimized labor costs and optimized resource allocation. The AI system also played a key role in balancing workloads more effectively, thus reducing overtime expenses and employee fatigue.

Resource planning is a significant cost for manufacturers. Overstaffing can lead to increased labor costs without a corresponding increase in production, while understaffing can result in missed deadlines, overtime costs, and reduced output. AI-driven resource optimization ensures that manufacturers can strike the right balance, leading to cost savings and improved efficiency.

By analyzing historical data, such as past production schedules, order volumes, and worker availability, AI algorithms can predict future resource needs. These predictions can then be matched with current resource levels and skills to create optimized resource plans. Additionally, AI can factor in variables like seasonal demand fluctuations, employee leave, and training needs to refine these plans further.

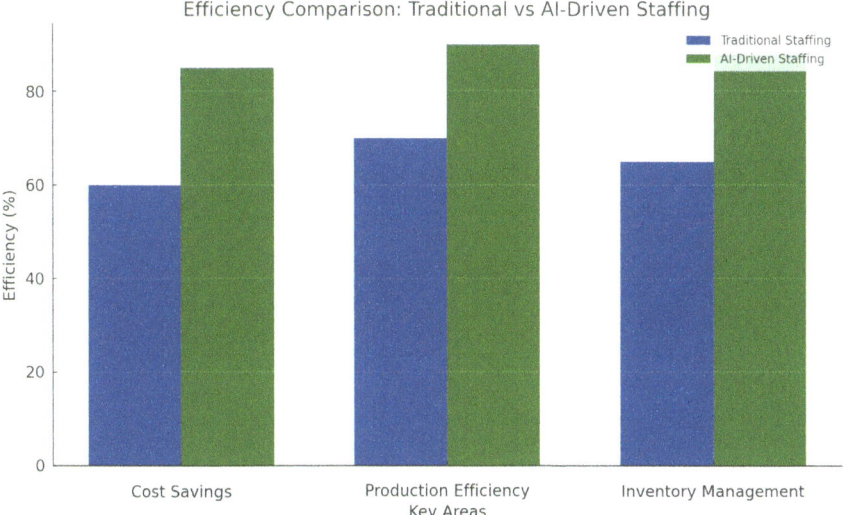

Fig. 3.7 AI-driven resources outperform traditional methods across cost savings, production efficiency, and inventory management. Data is for illustrative purposes only. Created by the authors in Python Notebook

3.4 Key Takeaways

- Machine learning (ML) algorithms analyze vast amounts of production data to predict equipment failures, reducing unplanned downtime and facilitating proactive maintenance.
- Advanced ML and deep learning techniques are used to detect defects with high accuracy, automate quality control processes, and ensure that only products meeting high standards reach consumers.
- AI-driven tools, including simulations and real-time data analysis, help in identifying bottlenecks and optimizing process flows, significantly increasing overall production efficiency.
- AI enables more precise resource planning and allocation, reducing labor costs by aligning workforce skills with production needs and adjusting resource levels dynamically based on predicted demand.
- By segmenting data and applying predictive models, manufacturers can forecast future disruptions and inefficiencies, allowing for data-driven decision-making that enhances operational strategies.
- AI applications excel in identifying complex patterns within large datasets, which is vital for spotting inefficiencies and potential improvements in manufacturing processes.
- Simulations powered by AI provide detailed modeling of manufacturing environments, enabling manufacturers to experiment and optimize processes without disrupting actual operations.

- ML algorithms assist in optimizing energy consumption and other resources, contributing to more sustainable manufacturing practices and compliance with environmental standards.
- AI's capability to continuously analyze equipment performance data leads to improvements in Overall Equipment Effectiveness (OEE), thus enhancing the reliability and performance of manufacturing operations.
- AI's role extends beyond manufacturing to assist in product design and facility planning, driving innovation by generating optimized designs and layouts based on real-world data and performance metrics.

Glossary

Generative Adversarial Networks (GANs) A type of AI model consisting of two neural networks that compete to produce new, synthetic data.

Generative AI AI systems capable of creating new content, designs, or solutions based on learned patterns and input data.

Large Language Models (LLMs) Advanced AI models trained on vast amounts of text data, capable of understanding and generating human-like text.

Lean Manufacturing A systematic method for waste minimization within a manufacturing system without sacrificing productivity.

Low-code AI AI development platforms that allow users to create applications through graphical user interfaces instead of traditional hand-coded computer programming.

MAPE-K Model A reference model for adaptive systems that stands for Monitor, Analyze, Plan, Execute, and Knowledge.

Overall Equipment Effectiveness (OEE) A standard metric in manufacturing that measures equipment efficiency based on availability, performance, and quality.

Partial Least Squares (PLS) A statistical method that finds a linear regression model by projecting predicted variables and observable variables into a new space.

Principal Component Analysis (PCA) A statistical technique used to reduce the dimensionality of large datasets while preserving important information.

Variational Autoencoders (VAEs) A type of generative model that learns to encode data in a lower-dimensional space and generate new data from this space.

References

1. Rapp, K. (2022). Artificial intelligence in manufacturing: Real world success stories and lessons learned [Internet]. *NIST*. Access date November 21, 2024. https://www.nist.gov/blogs/manufacturing-innovation-blog/artificial-intelligence-manufacturing-real-world-success-stories

2. Birlasoft. (2021). 17 remarkable use cases of AI in the manufacturing industry [Internet]. *Birlasoft*. Access date November 21, 2024. https://www.birlasoft.com/articles/17-use-cases-of-ai-in-manufacturing
3. NVIDIA. (2024). AI in manufacturing [Internet]. *NVIDIA*. Access date November 21, 2024. https://www.nvidia.com/en-us/industries/manufacturing/
4. Global AI implementation benefits in organizations. 2017. [Internet]. *Statista*. Access date November 21, 2024. https://www.statista.com/statistics/787094/worldwide-artificial-intelligence-implementation-in-companies-benefits/

Chapter 4
Intelligent Supply Chain

Abstract This chapter explores the use cases and applications of AI and ML in optimizing supply chain processes, enabling efficiency and agility across interconnected workflows. It discusses the use cases of ML in demand forecasting, contrasting it with traditional methods limited by rigidity and scale. The chapter includes discussion on AI-driven inventory optimization, which utilizes advanced algorithms to refine stock levels, placement, and replenishment strategies. Dynamic replenishment frameworks connect real-time consumption signals to automated production, mitigating risks of stockouts or overstocking. AI's application use cases in logistics are discussed for route planning, delay prediction, and contingency planning. Models such as LSTM networks improve traffic forecasting, while reinforcement learning optimizes delivery routes and schedules. Predictive analytics enable businesses to anticipate logistics delays and adapt proactively. The chapter includes detailed discussion on anomaly detection, root cause analysis, and KPI monitoring, showcasing AI's ability to identify disruptions and enhance decision-making. In addition, the chapter explores generative AI's potential in procurement, automating data-intensive tasks and refining strategic operations. Real-life implementations, including predictive cargo management and risk mitigation in logistics, demonstrate AI's practical impact. This chapter provides a comprehensive view of how AI and ML reshape supply chain resilience and performance.

Keywords Machine learning for demand forecasting · AI-driven dynamic inventory optimization · Logistics route optimization · Anomaly detection in supply chain · Generative AI for procurement · AI-powered supply chain management

Whereas Chap. 3 focused predominantly on enhancing analytics for individual manufacturing functions through AI/ML, this chapter progresses to process-wide integrations across intricate supply chain workflows. Manufacturing and supply chain processes prove deeply interdependent—from sourcing materials to fulfilling customer orders. Hence, optimizing these connections through data and AI promises tremendous efficiency gains.

This chapter starts with examining machine learning approaches for accurate demand sensing, contrasting these responsive models against rigid statistical methods unable to adapt to market fluctuations. By continually ingesting signals from sales data, events, and even social media, ML forecasting promises unprecedented agility.

Building on demand projections, AI optimization of inventory positioning strategies is discussed next. This includes guiding warehouse stocking quantities, safety stock buffers, and advanced shipment coordination to balance customer service with cost efficiency. AI also enables dynamic inventory replenishment by linking consumption alerts to automated production changeovers.

The chapter then explores AI-powered logistics route optimization and machine learning predicting likely shipping delays. This facilitates responsive scheduling, transportation mode shifting, and contingency planning—granting supply chain leaders tools to navigate market turbulence.

Finally, applications like leveraging machine learning for detecting order/ inventory outliers, AI-root cause analysis, monitoring KPIs, and the rise of generative AI to transform procurement workflows are discussed.

4.1 AI-Driven Optimization in Demand Forecasting, Inventory, and Replenishment

Accurately predicting customer demand and optimally planning inventory to align with forecasts remains an enduring challenge across complex global supply chain networks, potentially impacting revenues, customer service, and working capital. However, advancements in artificial intelligence and machine learning offer immense potential for enhancing traditional statistical demand sensing and inventory positioning approaches plagued by inadequate responsiveness and scalability. This section analyzes crucial applications of AI methodologies across the dynamic interplay among demand analysis, inventory planning, and replenishment systems.

In 2020, the COVID-19 pandemic led to a situation where 28 percent of retailers experienced shortages and instances of out-of-stock items. As of 2020, 64% of retailers worldwide adapted their supply chain for e-commerce as a result of the COVID-19 pandemic [1]. The topics discussed below focus on employing ML-based customer demand forecasting, AI-driven inventory optimization, and replenishment to address these challenges. These strategies are designed to effectively anticipate demand shifts and strategically manage inventory, thereby minimizing the impact of similar disruptions in the future.

4.1.1 ML-Based Demand Forecasting

Accurate demand forecasting serves as the crucial starting point for the efficient execution of many key supply chain activities, including warehousing, logistics, pricing strategies, financial planning, and more. Demand forecasting is tough, and most efforts made presently yield disappointing accuracy with noticeable prediction errors.

Existing legacy methods cannot easily identify patterns in-demand data, and their restricted capability to grasp the root causes of demand changes makes the volatility appear worse than if drivers were well understood. For example, conventional time-series statistical techniques like the Autoregressive Integrated Moving Average (ARIMA) model rely on analyzing historical order data, and because of the manual-intensive nature, there persists biases and poor productivity among planners.

Rather than appearing as a logical series of numbers, in today's business environment, demand more often seems like a pattern of partially constrained chaos. Demand is increasingly influenced by multiple internal and external factors that drive it up and down in ways that can't be understood by simply looking at a historical time series of aggregated demand buckets. Instead, demand should be viewed as being driven by a complex series of indicators that can be nearly impossible to manage with traditional forecasting algorithms.

The confluence of exponentially increased computing power and intensifying market volatility has established crucial prerequisites for the wider adoption of machine learning (ML) to drive predictive analytics across modern supply chain.

The "priority rating" column in the table provides indicative rankings to guide the usage and integration priority for each data source when building machine learning-based demand forecasting models.

4.1.1.1 Example Use Cases

The examples of machine learning-based demand forecasting models are outlined in the table below with the assigned priority levels:

High Priority:

- The data source exhibits a strong correlation with demand patterns
- Should be proactively incorporated as a core model input.

Medium Priority:

- The data has a moderate demand prediction signal
- Merit in selective usage; cost–benefit tradeoff evaluation needed.

Low Priority:

- Data poised marginal demand forecasting value
- Significant integration effort unlikely to be justified.

Situational Priority:

- Relevance contingent on product type or operating contexts
- Adopt if category steeply impacted by data factor.

In Table 4.1, two types of data are referred to—structured data, which is easily interpreted, and unstructured data, which needs more processing. There is an inherent trade-off between how readily these data types can be analyzed versus the richness of insights they may contain.

Structured data type is highly organized and conforms to a predefined data model, allowing for efficient analysis. It is stored in tables, relational databases, and other formats that enable easy searching. Examples of structured data that lends itself to quantifiable analysis include numbers, names, and dates.

On the other hand, unstructured data lacks conformity to a strict structure. However, this enables flexibility in the variety of qualitative data it can represent, from written language to images and video. While often containing some structured elements like dates or figures, unstructured data is predominantly loose text or multimedia difficult for systems to interpret without additional processing. Common examples ranging from text-heavy to rich media include emails, social

Table 4.1 Mapping between data source of demand forecasting and priority ranking

Data source	Examples	Priority rating	Data type
Past financials and sales history	Sales volumes and Revenue and profit metrics Seasonality patterns	High	Structured
Marketing polls	Purchase intent surveys, Voice of customer data and A/B testing performance	Medium	Unstructured
Macroeconomic indicators	GDP and employment growth projections, Consumer confidence indices Inflation rate changes	Low	Structured
Social media signals	Product buzz and mentions, Reviews and ratings spikes and Follower growth tracking	Low	Unstructured
Weather forecasts	Temperature/ precipitation outlooks and Extreme weather events predicted	Situational	Structured
News on local events	Sports events and festivals scheduled Infrastructure projects awarded	Situational	Unstructured
Competitor activity	Competitor pricing changes Promotional calendars Product feature releases	High	Unstructured

posts, photos, videos, and PDFs. Unstructured data requires cleansing and annotating before yielding meaningful patterns.

The priority levels provide a guideline on which demand indicators serve as the most material or additional drivers for enhancing the predictive power of ML models across different applications. The rankings signify effort-value tradeoffs to optimize model performance without overengineering.

ML Differentiator in Demand Forecasting—What Can It Do?

A key ML differentiator lies in automatically detecting intricate mathematical patterns within large, multidimensional datasets using sophisticated algorithms for continuous self-learning—contrasting with rigid statistical formulae reliance. Instead of humans specifying exact functional relationships or coding software rules, ML models independently recognize latent patterns and complex demand interconnections.

ML uniquely equips systems to ingest information volumes much beyond human-coded analytics capacities. Rather than sparse periodic analysis done conventionally, ML engines retrain operational models daily or even hourly using the latest integrated signals capturing market fluctuations. Thereby, forecasts continually adapt as conditions dynamically evolve versus static snapshots produced previously.

Volatilities stem from causes like supply crunches, new competitors in channels, promotions going viral on social media etc., which blindside legacy systems that can be continually routed as inputs to responsive ML models.

For example, Machine learning can interpret the effect of stimuli (such as trade promotions and advertising) and demand indicators (such as social media activity) originating from each distribution channel. As information proliferates or increases rapidly in numbers, the data concerning these causes and demand indicators become both more accessible and more manageable over time. Machine learning systems, therefore, can integrate and usefully model these important new data sources, including detailed market data, machine telemetry, and social media feeds, in ways that are simply not possible with legacy planning systems.

In essence, machine learning brings two crucial aspects—automated detection of complex data relationships and relentless model retraining ability—to transform dynamic signal capture. The mathematic and analytic heavy lifting embodied enables unlocking value from exponentially growing data at a pace aligned to today's business speed.

High accuracy of demand projections is important for supply chain forecasting systems. AI and ML-powered approaches assist enterprises to attain enhanced data-driven precision. As per current surveys, an overwhelming 80% of companies have embraced or intend to welcome machine learning and AI adoption into their forecasting processes, gaining competitive advantage through increased productivity.

Over 20 years back, retail major Nike invested $400 million to upgrade their demand forecasting systems. However, due to insufficient testing and poor data integration, they faced disastrous stockouts, losing $100 million instead. The deployed systems had bugs and couldn't smoothly exchange data across platforms, generating unreliable demand predictions.

This highlights two key success factors modern AI-based solutions mandate—strong data integrations for seamless inputs across sources and rigorous iterative testing to validate functionality and output accuracy before widespread usage. As Nike discovered through their complex rollout, unless backed by precise, interconnected data flows and methodical forecast reliability assurances, even extensive investments may severely backfire.

4.1.2 AI-Driven Inventory Optimization

Inventory optimization refers to using artificial intelligence and machine learning techniques to optimize decisions around inventory planning and positioning across a supply chain distribution network.

Demand forecasting, discussed in the earlier section, refers to the process of predicting and forecasting expected customer demand for products over a future time period. It focuses on answering the question—"How much demand will we have for different items?"

In contrast, inventory optimization and placement refer to decisions regarding—

- Target stock levels for the predicted demand.
- Positioning inventory geographically across warehouses, stores, distribution centers, etc.
- Managing shipment quantities across facilities.

Essentially, it covers the planning aspect of determining optimal inventory deployment strategies and responsive positioning tactics to best satisfy the demand forecasts generated.

While accurate demand forecasts serve as a crucial input, inventory optimization leveraging AI/ML goes beyond that to solve questions like:

- How much safety stock do we need?
- How do we distribute items across fulfillment centers to minimize shortages?
- How can we swiftly respond to demand upticks for efficiency?

In summary, demand planning focuses just on projections, while inventory optimization determines operational strategies and tactical plans based on those demand signals. AI and advanced analytics techniques help drive improvements in both areas to boost overall supply chain performance.

The scope of inventory level optimization spans both short and long-term strategic decisions around best bets for total stock amounts, their deployment across the supply/logistics network, and exchange locations. It also continually maintains responsiveness to demand changes through prompt inventory movement and positioning adaptations in the short run.

This encompasses both—

Strategic supply planning: Determining optimal overall stock levels, placement strategies, and inventory targets from a system-wide perspective. This tends to involve relatively longer-term planning horizons.

Short-term dynamic positioning: Figuring out daily/weekly shipment quantities across warehouses, adjusting regional distribution center inventory mixes, coordinating store-level stocks, etc. This represents more granular, responsive inventory positioning based on short-term signals like demand upticks or events.

A range of AI and machine learning techniques exhibit potential for optimizing supply chain inventory positioning and planning. Table 4.2 summarizes the key categories of models explored, how they work, as well as their respective strengths and limitations:

Table 4.2 ML model and their contributing factors in inventory optimization

ML model	How it works	Contributing factors in inventory optimization	Limitations
ML-based supervised learning techniques 1. Ensemble Methods 2. Deep learning 3. Linear methods	• Predictive model forecasts lost sales for placement decisions • Combination of simulation, optimization, and supervised learning models	• Leverages pattern recognition capabilities • Balances benefits of constituent techniques • Tailored to each use case and application, identify the technique with the best performance	• Susceptible to bias • Black box limitations • Architectural complexity • Computing resources • Implementation challenges
Genetic algorithms	Mimics natural selection to iterate inventory allocation combinations to optimize service level and cost metrics	• Handles constraints well • Identifies unexpectedly optimal solutions that may not be intuitive to humans	Slower than other techniques
Reinforcement learning	The inventory agent model learns ideal stock positioning policies via simulation to maximize reward functions around service, cost, etc.	• Highly customizable AI model • Optimal behaviors aligned with business objectives	• Simulation setup complexity • Many trainings iterations • Implementation challenges
Optimization solvers	Mathematical programming methods to optimize inventory levels under demand uncertainty	Quickly yields optimized allocation recommendations	• Limited uncertainty adaptability • Prone to overfitting

4.1.3 Dynamic Inventory Replenishment Based on AI

The responsiveness of inventory systems further relies on linking updated demand forecasts to supply and production signals through dynamic replenishment frameworks. Legacy ERP systems lack real-time synchronization of depletion signals across distribution nodes to activate timely restocking. AI again overcomes latency and scale limitations by automatically triggering supply workflows based on predictive insights.

For example, by continually processing sales data and historical inventory data, AI algorithms can rapidly detect patterns and cluster stores based on consumption profiles. Store clusters are automatically mapped to regional distribution hubs, then multi-level inventory optimization to align hub and store stocks with predicted demand. The AI algorithms likewise tie warehouse withdrawals to signals for transportation mode shifting or production changeovers.

Extreme automation also enables potentiating emerging rapid fulfillment models like flash retailing. Hub watcher algorithms localized to regional trends ensure prompt identification of spikes or stockouts to dispatch via distribution center in hours rather than days. AI is essential for inventory positioning to catch up to the pace of integration across multiple channels.

Blockchain smart contracts allow for trusted transactions between parties without central authority through automated self-enforcement of agreements encoded directly into the blockchain network. Blockchain smart contracts can use AI predictions of customer buying to improve supply chain management. Specifically, vendors could use AI forecasts of retailer demand to better manage inventory. This can smooth fluctuations in orders from retailers. The repeatability of machine learning forecasting, optimization, and monitoring processes also makes them ideal applications for robotic process automation to remove manual interventions and delays while adapting dynamically.

AI applications to streamline legacy forecasting, positioning, and replenishment systems require careful accounting for interacting processes and data connections. However, strengthening supply chain foundations using AI techniques discussed above may unlock immense potential for boosting productivity, agility, and growth.

4.2 Leveraging AI/ML for Logistics Route Planning and Predicting Delays

Logistics ecosystems prove intricately complex—encompassing interconnected decisions across transportation modes, warehouses, inventory positioning, and more. AI and advanced analytics introduce data-driven rigor to unravel and optimize these interdependencies while maintaining flexibility.

Logistics route planning is the process used to determine the optimal pathways to move goods through supply chain networks to minimize costs and delivery times.

Predicting delivery delays involves logistics analytics to estimate the likelihood and duration of late shipments.

4.2.1 AI for Logistics Route Planning

AI-powered route planning represents a cutting-edge technology that uses artificial intelligence and advanced algorithms to significantly enhance the efficiency of delivery routes.

Unlike legacy systems with fixed shipment logic, AI-based dynamic routing responds to dynamic conditions. It continuously synthesizes variables such as weather, traffic, warehouse workload, and asset availability and predicts local demand upticks to chart optimal routes.

AI-powered route planning is reshaping how businesses plan and execute their delivery operations in the ever-changing landscape of transport networks. Integrating diverse factors and data ensures transportation firms formulate optimal routes, reducing operating costs and optimizing resource utilization.

In supply chain route planning, a factor refers to a variable element related to the real-world routing environment and network conditions that influence delivery efficiency, capacity utilization, or other key performance metrics. These factors impose constraints that algorithms must account for when computing optimized routes and logistics plans.

In Table 4.3, the mapping between the route planning approach and AI route planning algorithms and AI model are shown.

Effective logistics route planning requires finding a balance between the efficient scaling of complex evaluations beyond human capabilities through AI and the implementation of checks to ensure viability under diverse conditions. The below detailed analysis of the route planning approaches emphasizes the interconnection of automated intelligence and practical human wisdom to derive the best result out of the AI system.

4.2.1.1 Traffic Prediction Using Long Short-Term Memory (LSTM) Networks

Long Short-Term Memory (LSTM) networks are exceptional in identifying useful patterns in large time-series data. This allows reliable forecasting of delays across routes, enabling proactive logistics planning. While LSTM-based traffic predictions are generally reliable, there are instances when they may fall short. Therefore, it is important to validate predictions before complete reliance. Conducting thorough testing to pinpoint limitations, retraining models for new regions, and incorporating mathematical checks can help maintain the robustness of the solution.

Table 4.3 Route planning approach and AI route planning algorithms, AI model

Route planning approach	Algorithm examples	AI model	Description
Traffic prediction	Long Short-Term Memory (LSTM) networks	Deep neural network	Uses historical and real-time data to forecast congestion hotspots
Delivery time windows	Constraint programming, reinforcement learning policies	Mathematical optimization + AI automation	Incorporates node-specific delivery window constraints
Location sequencing	Travelling Salesman Problem (TSP) solvers, reinforcement learning route plans	Combinatorial optimization + Machine learning simulation	Considers proximity, accessibility etc., to optimize delivery sequence
Vehicle capacity	Knapsack optimization, reinforcement learning agent	Mathematical optimization + Machine learning simulation	Matches assigned routes to Commercial Vehicle capacity
Historical data dependency	Clustering models	Unsupervised machine learning	Continually learns from traffic patterns, constraints, etc.

4.2.1.2 Delivery Time Windows

Delivery Time Windows Optimization leverages historical delay data to pad routes with buffers that minimize late deliveries. Though padding helps reliability, over-padding routes bloat costs if buffer durations misalign with emerging norms. Continual assumptions validation through data audits, adapting buffers responding to updated norms, and dynamic tuning aligned to the latest conditions balances optimization with on-time delivery guarantees.

4.2.1.3 Location Sequencing

Optimized location sequencing reduces time and mileage compared to manual routing. However, it is oversimplified by neglecting specific node considerations, such as treating an urgent hospital shipment with the same priority as a furniture delivery. To address this, configuring hierarchical group rules (e.g., prioritizing city over rural areas) and internal priority indices enables the incorporation of constraints like regional emergencies and security checks. Frequent retraining on new nodes continually enhances recommendations. This approach combines mathematical precision with human wisdom to guide route planning.

4.2.1.4   Vehicle Capacity

The underlying mathematical techniques efficiently optimize the allocation of shipments across vehicles, ensuring a balance in utilization, cargo constraints, densities, and more. However, the abstraction of real-world loading intricacies poses the risk of overloads and stability issues without appropriate validity checks. To mitigate this, simulating actual weight distributions and conducting pilot validations at dispatch helps prevent such physical oversights. In essence, the combination of rigorous optimization and practical verification enhances planning precision while perpetually automating complex evaluations.

4.2.1.5   Historical Data Dependency

Continually improving recommendations by identifying geo-correlations across nodes from accumulated data is great. However, the occasional data collection and privacy rules that limit sharing can create biases and overfitting risks. To tackle this, it's crucial to rigorously check data quality by cleaning anomalies and running simulations.

4.2.2   ML for Predicting Logistics Delays

Supply chain face immense pressure from logistics delays and disruptions, estimated to cause nearly $4 trillion in lost revenues since 2000, making Just-in-Time (JIT) manufacturing a challenging proposition. Monitoring supplier operations is crucial to building resilient and ethical supply chain. The next-gen supply chain system uses AI technology to continuously monitor suppliers and logistics situations, ensuring compliance with responsible sourcing policies and ESG standards. The AI-powered tracking and analysis provide real-time visibility into suppliers so procurement teams can make data-driven decisions.

   Similar to a silent storm brewing, logistics delays, and disruptions have become inherent in today's intricate supply chain, posing challenges for traditional planning methods. Whether it's the Suez Canal blockage, a single event that significantly hindered billions in daily trade, or occurrences like wars, natural disasters, and the ongoing impact of the COVID-19 pandemic, the spectrum of disruption sources is extensive. Logistics delays adversely impact operations and customer service in supply chain. However, shipping via complex logistics networks makes delays and uncertainties inevitable. Manual track-and-trace processes also prove expensive and offer limited functionality.

   By intelligently synthesizing signals from historical transport patterns across diverse shipping routes, carriers, and geographies, ML models can uncover crucial interrelationships. Factoring key variables like weather forecasts, traffic, and capacity and analyzing their probabilistic impact allows for predicting likely delay risks.

Rather than guessing based solely on recent point delays, machine learning takes a systemic approach. It quantifies the cumulative effect of relevant indicators such as port congestion, pandemic restrictions on operations workforce, and more to forecast impacts on delivery lead times even over longer advanced time horizons.

Local tailoring proves critical to precision—models tuned to air freight parameters predict starkly differently versus seaports or road carriers. Isolating peculiar nodes also improves granularity. This allows model outputs to project delay likelihoods for specific shipping lanes or even transport batches.

The quantifiable risk alerts with lead time allow sensitive product shippers to plan production cycles accordingly or switch transit modes accordingly. Even scheduling shipment transfers proactively through predicted transport bubbles minimize revenue risks and inventory pileups.

Organizations are increasingly adopting AI-powered predictive analytics to proactively navigate and minimize the impact of risks around logistics delays. Upon flagging risks, AI can simulate mitigation strategies, providing robust and economical responses—such as adjusting shipment batch sizes or diversifying the supplier mix. Successful organizations have optimized inventory and proactively adjusted stock levels based on AI forecasts, especially during volatile demand swings.

Through continuous analysis of historical data, AI models can identify predictive patterns associated with potential disruptions. The use of cloud platforms expedites the development of custom algorithms tailored to supply chain using enterprise data.

Below is an example of two successful use cases—

- Real-time views of supply chain disruptions: A leading logistics provider leveraged predictive analytics for providing near real-time global visibility and precise classification of disruptions, significantly enhancing proactive risk management and resilience across its global supply chain.
- Predicting cargo no-shows: A premier airline leveraged machine learning models to predict when shipments fail to deliver with at least 90% accuracy, improving cargo space utilization and fuel efficiency while mitigating revenue losses.

4.3 Detecting Outliers to Identify Root Cause and Monitoring KPIs

Outlier detection (also known as anomaly detection) is the process of finding data objects with behaviors that are very different from expectations. Such objects are called outliers or anomalies.

The most interesting objects are those that deviate significantly from the normal object. Outliers are not being generated using the same mechanism as the rest of the data.

The below section outlines how, by leveraging advanced machine learning to detect outliers, using artificial intelligence to dig deep into the roots of supply chain issues, and closely monitoring Key Performance Indicators (KPI) for timely alerts on any deviations, a robust and optimized supply chain system can be built.

4.3.1 ML-Driven Outlier Detection in Orders, Inventory, and Shipments

Machine learning models can analyze historical supply chain data—including past order volumes, inventory levels, and delivery rates, to forecast normal ranges and detect unexpected variances indicative of disruptions.

Various ML techniques, such as supervised, semi-supervised, and unsupervised methods, including neural networks, Support Vector Machine (SVM), clustering algorithms, etc., are employed to identify intricate outlier characteristics.

For example, clustering algorithms can analyze order volumes across regions and channels over time to define expected variation bands. By comparing incoming sales with these dynamic clusters, any notable spikes or dips beyond the learned boundaries can signal potential anomalies in demand for further investigation.

Likewise, custom neural networks and deep learning models can ingest inputs such as order, inventory, and logistics datasets. These networks can discover latent patterns, simulate expected ranges, and detect points straying beyond modeled normal behavior, indicating supply chain exceptions.

The most suitable approach depends greatly on data types, volumes, and infrastructure availability, among other factors. Testing combinations of supervised and unsupervised ML techniques allows tuning the best approach over time. Focusing outlier detection initiatives specifically on machine learning provides cutting-edge capabilities for unlocking insights hidden deep across fragmented supply chain data.

4.3.2 AI-Driven Root Cause Analysis of Supply Chain Data

Once an ML model flags potential outliers discussed above, additional logic can work to formulate a hypothesis about what core factors or root causes led to the anomaly. For example, neural networks incorporating graph theory leverage supply chain topological maps, connecting related nodes like suppliers, manufacturing centers, warehouses, and customer locations.

By analyzing the relationships and interdependencies both upstream and downstream from outlier data points, the AI can trace along these mapped connections to highlight possible cause-and-effect relationships. This root cause analysis guides investigation around not just the abnormal data itself but also interconnected nodes involved in the overall outlier event.

4.3.3   Monitoring KPIs to Alert on Deviations

Patterns in data resulting from events like the COVID-19 pandemic, such as significant drops in order intakes or extensive inventory backlogs, might persist as "new normal" for a considerable period rather than sporadic outliers limited to a single company or subset of items. To accurately model these mass disturbance events, it is important to establish appropriate thresholds or KPIs for outlier detection by incorporating the external context of such disruptions.

Setting up consistent monitoring of operational key performance indicators (KPIs) tied to supply chain processes allows employees to stay on top of potential disruptions or underperformance. Rather than randomly inspecting datasets, organizations can identify their most critical metrics, e.g., order fill rate, days of inventory outstanding, delivery appointment adherence, and configure persistent auditing.

Static rules can alert if a KPI crosses a defined threshold, but ML may more effectively track historical patterns to trigger smarter notices of more serious deviations. Building a workflow for regular automated KPI report generation and distribution pushes supply chain visibility to key leaders and frontline decision-makers. Additionally, a layered alert severity structure can escalate notifications appropriately based on the scale or sustainment of KPI anomalies.

Proactively monitoring supply chain KPIs through automated dashboards and smart notifications removes reliance on individuals manually tracking disparate datasets. Leveraging the above techniques, enterprises can cut through growing data complexities to stay continuously informed, quickly investigate deviations, and orchestrate interventions—ultimately driving more resilient operations.

4.4   Gen-AI-Powered Virtual Assistant for Smart Procurement

In procurement, Generative AI is rapidly gaining traction as a technology with vast untapped potential. The emergence of Large Language Models (LLMs) can play a significant role in Source-to-Pay, and this is just the beginning of what's possible. Procurement experts are encouraged to adopt the capabilities of Generative AI, especially its no-code applications, to innovate and develop custom solutions tailored to their specific needs.

Large language models (LLMs) are a special type of generative AI that transforms the way text data is handled in procurement. The LLM models are designed to understand, interpret, and generate human-like text, making them incredibly powerful tools for a variety of complex tasks. In procurement, LLMs can be used to analyze vast amounts of text data, streamline communication processes, and aid in strategic decision-making. Their ability to process and generate human-like text quickly and accurately makes them invaluable for tackling the nuanced and often text-heavy elements of procurement tasks.

4.4.1 Why is Generative AI Critical for Procurement?

Generative AI can have wide-ranging applications in procurement due to its ability to address key challenges faced in modern procurement environments. As procurement processes become increasingly complex with escalating data and task volumes, shrinking timeframes for strategic planning, and expanded responsibilities amidst constrained resources, Generative AI can significantly enhance the efficiency of the procurement process. It streamlines data management, enhances strategic planning within tight deadlines, and automates routine tasks, enabling procurement professionals to focus on strategic priorities. Additionally, its capability to tailor communication for diverse stakeholders ensures smooth and effective procurement processes. These applications demonstrate that Gen-AI can play an indispensable role in modern procurement strategies -

Escalating data, information, and task volumes:

In the modern procurement process, there's a significant increase in data, information, and task volumes. This escalation is due to the vast digitalization of procurement processes and the increasing complexity of supply chain. Procurement professionals now face the challenge of managing and interpreting this growing data deluge, necessitating more advanced tools and methodologies to efficiently handle the surge.

Generative AI can streamline data analysis and management. It can automate the processing of large datasets, enabling quicker and more accurate insights, thereby alleviating the burden on procurement professionals.

Shrinking timeframes for strategic planning:

The pace of business is accelerating, resulting in shorter timeframes for strategic planning in procurement. In order to meet the rapid market changes and evolving business needs, procurement professionals often utilize quicker decision-making with less time for thorough analysis. This puts pressure on procurement teams to be more agile and responsive in their strategic approaches.

Generative AI can significantly enhance strategic planning in tight timeframes. Rapidly analyzing trends and generating predictive models enables faster, data-driven decision-making, which is crucial for keeping pace with the rapid changes in the market.

Expanded responsibilities with constrained resources:

Procurement roles are expanding, encompassing more strategic responsibilities, often without a corresponding increase in resources. This expansion includes managing supplier relationships, ensuring sustainable practices, and mitigating risks. These growing responsibilities must be handled with existing or even reduced resources, making efficiency and prioritization key.

With expanded responsibilities in procurement, Generative AI offers a solution by automating routine tasks and providing intelligent insights. This allows procurement professionals to focus on strategic tasks, maximizing the impact of limited resources.

Communicating effectively to meet the diverse needs of all stakeholders:

Effective communication is crucial in procurement, especially given the diversity of stakeholders involved. This includes suppliers, internal teams, and external partners. The challenge lies in ensuring clear, consistent communication that addresses the unique needs and concerns of each stakeholder group, facilitating smooth and successful procurement processes.

Generative AI can aid in customizing communication to meet the needs of diverse stakeholders in procurement. Analyzing communication patterns and preferences can help tailor messages for clarity and effectiveness, ensuring smooth stakeholder interactions.

Procurement faces an ever-growing workload, marked by escalating volumes of data, an extensive array of stakeholders, and intricate strategies to put into action–all within the constraints of limited resources.

Transform the procurement strategy and operations with the power of artificial intelligence to automate time-intensive daily tasks such as research, analysis, content creation, modification, and summarization.

Gen AI will address these challenges to propel the Procurement function into a more efficient, effective, and strategic future.

4.4.2   Procurement Use Cases Powered by Gen AI

In this section, we explore various practical applications of Generative AI in procurement, demonstrating the various innovative ways it can be integrated into different aspects of procurement processes. These applications range from conducting supplier research and legal assistance to providing category and market insights.

Gen AI also enhances communications through content creation and proofreading and assists in generating improvement plans for suppliers. Additionally, it offers the unique capability to interact with and extract vital information from documents, such as supplier reports and contracts, demonstrating its versatility in procurement. Here are some of the use cases -

@ @**Supplier Research:**

Utilize Gen AI to gather essential information about both existing and potential new suppliers, including details about their products and services.

Legal Assistance:

Leverage Gen AI to quickly generate contract summaries or draft specific clauses, ensuring key elements are comprehensively covered.

Category and Market Insights:

Employ Gen AI to conduct in-depth research and provide concise summaries of key trends within specific procurement categories.

Enhanced Communications:

Benefit from Gen AI's capabilities in content creation and proofreading to refine and enhance communication materials as per specific instructions.

Improvement Plan Generation:

Use Gen AI to analyze supplier performance, summarize findings, and suggest actionable steps for structured improvement plans.

Interactive Documents:

Engage Gen AI to extract relevant information, data, and insights rapidly from various documents, such as supplier Environmental, social, and governance (ESG) reports, contracts, and financial filings.

4.4.3   Gen-AI-Powered Virtual Agents for in Warehouse

The integration of voice assistants in warehouses can optimize logistics and streamline operations in AI-Driven Manufacturing Processes. These voice-activated systems are enhancing the efficiency of warehouse tasks, from inventory management to order picking, by offering hands-free, real-time assistance.

Voice assistants, akin to the ones we use in our daily lives, like Amazon's Alexa or Google Assistant, are AI-driven systems designed to understand and respond to voice commands. In the warehouse setting, these assistants are tailored to cater to specific tasks associated with logistics and inventory management. They interact with workers, guiding them through tasks, offering real-time data, and ensuring accurate and efficient operations.

Warehouses are the backbone of a supply chain and are a hub of a multitude of activities. The efficiency and accuracy of operations here directly impact the broader manufacturing process. Traditional methods, often reliant on manual data entry or handheld devices, can be cumbersome and prone to errors. Voice assistants, with their hands-free operation, ensure that workers can perform tasks without interruptions. They reduce the chances of errors, ensure safety by allowing workers to maintain focus, and significantly speed up processes.

Voice assistants, integrated with the warehouse management system, offer real-time data access. Workers can ask about inventory levels, locate products, or even get guidance on the optimal route within the warehouse. These systems can also be integrated with other smart devices, ensuring a seamless flow of information. For tasks like order picking, voice assistants provide step-by-step guidance, ensuring accuracy and efficiency.

Table 4.4 provides an overview of the nature, comparison with traditional methods, system integration, and specific applications of voice assistants in warehousing, along with their benefits and practical applications.

Example Use Cases:

Table 4.4 Overview of voice assistants' role and impact in warehousing

Topic	Description	Solution strategies and applications
Nature of voice assistants	AI-driven systems understand and respond to voice commands	Hands-free operation, real-time data access
Traditional methods versus voice tech	Traditional methods use manual data entry or handheld devices	Voice tech ensures accuracy, speed, and safety
Integration with systems	Voice assistants connect with warehouse management systems and other smart devices	Seamless flow of information and accurate data access
Applications in warehousing	Used in pick-to-voice systems, inventory checks, and safety protocol guidance	Enhances order accuracy, reduces stock issues, and ensures safety

Pick-to-Voice Systems

Voice assistants have found a significant application in warehouses through pick-to-voice systems. These systems provide workers with voice commands that guide them directly to the precise location of items within the warehouse. This technology ensures accurate order fulfillment by minimizing the risk of human error in item selection. Additionally, it reduces the time spent searching for products, thereby increasing operational efficiency. The integration of voice assistants in this manner significantly improves warehouse management, leading to improved productivity and order accuracy.

Inventory Checks

Voice assistants can improve the way inventory checks are conducted in warehouses. Instead of workers manually checking or scanning items, they can simply ask the voice assistant about inventory levels. This ensures real-time access to data, significantly reducing the time spent on inventory checks. It also minimizes the chances of stockouts or overstocking, as the data provided by the voice assistant is accurate and up-to-date. This technology not only enhances efficiency but also contributes to better inventory management and control.

Safety Protocols

Voice assistants can play a crucial role in ensuring the safety of warehouse workers. They can provide guidance on safety protocols, ensuring that these procedures are followed meticulously. This reduces the chances of accidents or mishandling of goods, contributing to a safer and more efficient work environment. By reinforcing safety measures and providing real-time guidance, voice assistants can help create a culture of safety within the warehouse, protecting employees and assets alike.

4.5 Real-Life Examples of AI Use in Supply Chain

These real-world examples demonstrate how AI can revolutionize supply chain management and logistics in advanced manufacturing. By offering real-time visibility into disruptions and precise predictions of cargo no-shows, AI empowers manufacturers to preemptively handle risks, optimize resources, and enhance operational efficiencies. DHL's implementation of AI in its Resilience360 platform exemplifies how machine learning and natural language processing can enhance supply chain visibility and risk management, enabling proactive mitigation strategies against disruptions in logistics post-manufacturing.

> **Real-time Views of Supply Chain Disruptions**: DHL implemented AI into its Resilience360 platform to enhance visibility and risk management across global supply chain. Using machine learning and natural language processing, DHL improved the classification and prediction of disruptions, enabling proactive risk mitigation strategies. This resulted in heightened resilience against a range of challenges, including climate impacts, pandemic-related disruptions, incidents like cargo ship blockages in canals, geopolitical issues, and war situations [2].
>
> **Predicting Cargo No-Shows**: American Airlines utilized AI to predict cargo no-shows with 90% accuracy, using historical booking data and GPU-accelerated machine learning models. This enabled proactive management of cargo space, optimizing layout planning and fuel efficiency. The AI initiative also improved customer engagement through a fair booking policy, resulting in significant revenue savings and operational efficiencies [3].

4.6 Key Takeaways

- The adoption of Artificial intelligence is key in transforming supply chain management, from demand forecasting and inventory optimization to logistics and delivery. AI enables more accurate predictions and efficient operations, adapting in real-time to market and environmental changes.
- Machine learning models outperform traditional statistical methods by integrating diverse data sources such as sales data, social media, and external events, providing agility and precision in demand sensing.
- AI applications in inventory management not only predict demand more accurately but also optimize inventory placement and quantities across distribution networks, balancing cost efficiency with service levels.
- AI enhances inventory systems by linking updated demand forecasts to supply signals, enabling automated, real-time inventory replenishment and reducing risks of stockouts or overstocking.
- AI offers significant improvements in logistics by optimizing delivery routes and schedules based on real-time data inputs like traffic and weather conditions, which minimizes delays and lowers transportation costs.

- Machine learning models analyze various risk factors to predict potential delays in logistics, allowing businesses to proactively manage and mitigate the impact on the supply chain.
- AI technologies provide supply chain managers with real-time visibility and analytics, enabling quicker and more informed decision-making that aligns with current market demands and conditions.
- AI tools identify anomalies in supply chain data and help pinpoint underlying causes, facilitating timely corrective actions to prevent further disruptions.
- AI systems monitor key performance indicators in real time, helping organizations quickly identify and address deviations from expected performance levels.
- Large Language Models and other generative AI tools are revolutionizing procurement by automating data-intensive tasks, enhancing communication, and enabling more strategic decision-making.
- Voice assistants and AI-driven systems in warehouses improve efficiency, accuracy, and safety by supporting workers with hands-free operations and real-time guidance.
- The integration of blockchain technology and smart contracts in supply chain management ensures transparency and trust in transactions, enhancing the reliability and efficiency of supply networks.

Glossary

ARIMA (Autoregressive Integrated Moving Average) A statistical analysis model used for time series data to better understand the data or predict future points in the series.

Blockchain A distributed, decentralized, public ledger technology used to record transactions across many computers.

ESG (Environmental, Social, and Governance) A set of standards for a company's operations that socially conscious investors use to screen potential investments.

Just-in-Time (JIT) Manufacturing An inventory strategy that aligns raw-material orders from suppliers directly with production schedules.

KPI (Key Performance Indicator) A measurable value that demonstrates how effectively a company is achieving key business objectives.

Large Language Models (LLMs) Advanced AI models trained on vast amounts of text data, capable of understanding and generating human-like text.

LSTM (Long Short-Term Memory) A type of recurrent neural network capable of learning long-term dependencies, especially useful for time series prediction.

Pick-to-Voice Systems Warehouse management systems that use voice commands to guide workers through picking processes.

Smart Contracts Self-executing contracts with the terms of the agreement directly written into code.

Source-to-Pay The end-to-end process of sourcing goods and services, from identifying potential suppliers to paying for goods and services.

Support Vector Machine (SVM) A supervised machine learning model used for classification and regression analysis.

Traveling Salesman Problem (TSP) An optimization problem in computer science and operations research focused on finding the shortest possible route that visits each city exactly once and returns to the origin city.

References

1. Retail: Changes to supply chain due to coronavirus worldwide 2020 [Internet]. *Statista*. 2020. Access date November 21, 2024. https://www.statista.com/statistics/1143426/coronavirus-cha nges-to-supply-chain-retail-worldwide/
2. *DHL supply watch uses machine learning to mitigate supplier risks* [Internet]. Supply & Demand Chain Executive. 2017. Access date November 21, 2024. https://www.sdcexec.com/ software-technology/press-release/12337269/dhl-supply-chain-dhl-supply-watch-uses-mac hine-learning-to-mitigate-supplier-risks
3. Castro, N. (2020). *NVIDIA blogs: Quadro helps American airlines to improve weight distribution* [Internet]. NVIDIA Blog. Access date November 21, 2024. https://blogs.nvidia.com/blog/ame rican-airlines-data-science-workstations/

Chapter 5
Implementation Challenges and Solutions

Abstract This chapter addresses the multiple challenges faced by manufacturing organizations in implementing AI-driven processes and provides actionable solutions for overcoming them. Key areas of focus include handling large, diverse datasets, ensuring model explainability, embedding AI within legacy systems, and fostering cultural transformation to support AI adoption. The chapter introduces frameworks for scalable data infrastructure and robust data cleansing to manage the complexities of manufacturing data. It also explores advanced techniques such as LIME and SHAP for enhancing model interpretability and accountability, ensuring ethical AI practices. Solution strategies for integrating AI into existing workflows are covered in detail, including the use of MLOps for automating machine learning pipelines, APIs for seamless system interoperability, and containerized microservices for scalability. The importance of aligning AI initiatives with regulatory compliance and ethical considerations is emphasized, along with practical approaches for ongoing monitoring and adherence. The chapter also highlights the critical role of employee upskilling and leadership buy-in in driving successful AI transformation. Change management strategies and cross-functional collaboration are explored as enablers of organizational alignment. Illustrated through real-world use cases, this chapter equips manufacturers with end-to-end guidance to navigate AI implementation challenges and achieve sustained innovation and efficiency.

Keywords Scalable data infrastructure · Ethical AI practices · Model explainability techniques · MLOps · Regulatory compliance solutions · Change management strategies

As manufacturing companies embark on the journey of AI implementation, they invariably face common challenges around data, systems, accountability, and culture change. To showcase practical frameworks, organizations can adopt to effectively navigate these hurdles, let us walk through a representative case study of a fictional manufacturing firm called IntelliMachines Inc.

IntelliMachines Inc. is an ambitious manufacturing company keen to integrate AI and ML across its operations to enhance quality, efficiency, and competitiveness.

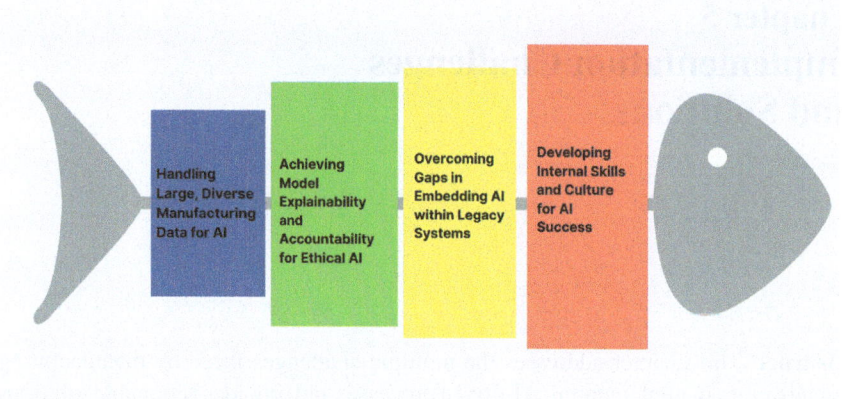

Fig. 5.1 AI transformation journey of IntelliMachines Inc. *Source* Created by the authors in Canva

However, transforming proven legacy processes using uncharted advanced technologies requires structured planning around priorities, investments, and execution strategies spanning tools, techniques, and talent.

Through the story of IntelliMachines' AI adoption journey across four critical focus areas, we illustrate solution frameworks manufacturing enterprises can replicate on their quest to become AI-driven powerhouses of the future. Beyond just showcasing AI use cases, we emphasize holistic planning to address process and cultural barriers that commonly derail manufacturing AI projects lacking structured roadmaps.

Story: The AI Transformation Journey of IntelliMachines Inc

Figure 5.1 displays a fishbone diagram providing a graphical overview of the key challenges IntelliMachines Inc. faced during their manufacturing industry AI adoption journey, along with corresponding solutions. The "head" of the fish structure represents the main transformation objective—to become an AI-driven organization. The major "bones" illustrate four key challenge categories.

Handling Large, Diverse Manufacturing Data for AI

As a first step in their AI journey, IntelliMachines faced challenges in managing their vast and varied manufacturing data. They established a scalable cloud infrastructure to aggregate data from sensors and machines on the factory floor. Their data engineers then developed standardized processes for cleansing and preparing the collated datasets. By handling missing values and anomalies, they ensured high-quality data for training AI models.

Achieving Model Explainability and Accountability for Ethical AI

Next, IntelliMachines focused on interpretability, documentation, and communication of their AI models' capabilities. Their data scientists integrated techniques like

LIME and SHAP to explain model predictions. By maintaining thorough documentation covering model development stages and performance benchmarks, they enabled auditing and incremental improvements. Overall, these strategies reinforced ethical accountability.

Overcoming Gaps in Embedding AI within Legacy Systems

To embed AI within legacy manufacturing platforms, IntelliMachines implemented MLOps automation for faster experimentation/deployment. With APIs and microservices, they enabled flexible integration. By assessing regulatory needs upfront and setting up ongoing compliance checks, they ensured AI innovations met quality and audit standards.

Developing Internal Skills and Culture for AI Success

Recognizing that technology alone cannot guarantee change, IntelliMachines invested in extensive AI/ML skills training and culture transformation programs. With personalized learning paths, they enabled employees to create an impact with AI tools. By securing leadership buy-in, promoting cross-team coordination, and celebrating small wins, they drove sustained momentum toward becoming an AI-driven, future-ready manufacturer.

In summary, by methodically overcoming challenges around data, models, systems, and skills, IntelliMachines succeeded in responsibly transitioning itself into an AI-driven industry leader. Their 360-degree execution approach balancing process innovation and cultural transformation became an inspiration for others to strive towards AI excellence.

By discussing IntelliMachines' systematic approach around these focus areas, we set the context for Chap. 4's detailed exploration of tangible strategies manufacturing firms can replicate to become AI trailblazers within their industry. Chapter 4 provides end-to-end guidance—spanning data, models, systems, and talent to overcome implementation barriers through structured frameworks for responsible and value-driven AI adoption.

5.1 Handling Large, Diverse Manufacturing Data for AI

According to an analysis by Fortune Business Insights, the Big Data Analytics sector shows significant growth potential, with a strong projected CAGR of 13.0% from 2024 to 2032. The market is expected to nearly triple in size from 2023 to 2032, driven by digital transformation, technological advancement, and increasing adoption across industries [1].

In the age of Industry 4.0, as AI-driven manufacturing processes become increasingly prevalent, the role of data becomes very important. However, the sheer volume, variety, and velocity of data generated in modern manufacturing environments present unique challenges.

Table 5.1 Strategies for managing and preprocessing manufacturing data for optimal AI application

Topic	Description	Solution strategies
Scalable data infrastructure	Infrastructure that can aggregate and process vast data volumes efficiently, being agile, flexible, and capable of handling data from diverse sources	Setting up cloud-based systems that scale with increased data from IoT devices on the factory floor
Cleansing and preparing heterogeneous data sets	Addressing missing values, anomalies, and biases in varied manufacturing data, ensuring its integrity and relevance for AI processing	Using data wrangling tools to standardize and normalize data from different sensors before feeding it into an AI model
Continuous monitoring and quality control	Implementing mechanisms to ensure data remains accurate, consistent, and up-to-date, reflecting the ever-evolving manufacturing environment	Setting up real-time dashboards that highlight anomalies in data input, allowing for quick corrections

5.1.1 Key Challenges for Manufacturers

Manufacturing industries, in their quest for efficiency and innovation, are generating vast amounts of data from a plethora of sources, ranging from Internet of Things (IoT) sensors on the factory floor to supply chain logistics. This data, while invaluable, is often varied, encompassing different formats, structures, and granularities. Managing and preprocessing this flood of data to make it suitable for AI applications is a monumental task.

The efficacy of AI-driven manufacturing processes hinges on the quality and readiness of the data fed into them. Properly managed and preprocessed data ensures that AI models are trained accurately, make precise predictions, and drive optimal manufacturing outcomes. Conversely, poorly managed data can lead to inefficiencies, errors, and missed opportunities.

Table 5.1 presents various strategies for managing and preprocessing manufacturing data to optimize AI applications.

5.1.2 Scalable Data Infrastructure for Aggregating Data

Scalable data infrastructure refers to the systems and processes that allow for the efficient collection, storage, and analysis of large volumes of data. In AI-driven manufacturing, this infrastructure is key for collecting data from various sensors, machines, and processes and then processing this data to derive insights that can improve manufacturing outcomes.

As industries move towards more automated and intelligent processes, the volume of data generated is immense. For instance, during the COVID-19 crisis, digital adoption surged, accelerating the need for modern data architectures. Leading AI adopters are investing even more in AI in response to these changes. However, despite its importance, many companies need help with foundational shifts necessary for innovation. Disparate data models and data quality issues can hinder AI development processes.

Moreover, traditional architecture design and evaluation approaches can paralyze progress. Organizations often get bogged down in overplanning, technology assessments, and stakeholder debates. This gridlock can be broken by rethinking modernization efforts and adopting practices that accelerate these efforts.

5.1.2.1 How to Implement a Scalable Data Infrastructure for AI Applications?

Establishing a scalable data infrastructure for AI applications involves strategic planning and the adoption of proven methodologies. The steps are discussed below:

Instead of starting from scratch, data and technology leaders can utilize reference data architectures that have been proven effective across various industries. These architectures offer agility, speed, flexibility, and innovation. For instance, a large German bank reduced its data-architecture blueprint definition time from three months to four weeks by leveraging a reference data architecture.

Instead of viewing data-architecture transformations as linear projects, organizations can focus on building a Minimum Viable Product (MVP) for specific use cases. This approach allows for faster results and iterative improvements based on user feedback. Educating business leaders about the limitations of legacy technologies and the advantages of modern solutions can facilitate smoother transitions. For instance, data lakes offer significant cost reductions and flexibility compared to legacy solutions.

Successful modernization requires an integrated team with a culture centered around data. This involves re-orienting the data organization towards a product and platform model and fostering a culture of continuous learning and innovation. Changing data architectures can be cumbersome. DataOps, which applies a DevOps approach to data, can automate many of these processes, allowing engineers to focus on code building. This approach can significantly speed up the deployment of new data models and pipelines.

Examples of scalable data infrastructure adoption:

A European fashion retailer streamlined its data architecture to provide data scientists with fast access to data, enabling them to personalize offerings across multiple channels effectively.

A pharmaceutical company used DataOps to test new biometric analytics models against standards, optimizing code reuse and reducing the time lag between model development and application.

As AI-driven manufacturing processes become the norm, having a scalable data infrastructure is not just an advantage but it's a necessity. By implementing the practices outlined above, organizations can ensure they are well-positioned to utilize the power of AI and drive innovation in their manufacturing processes.

5.1.3 Cleansing and Preparing Heterogeneous Data Sets

In AI-driven manufacturing processes, the availability of operational data plays a significant role. Data serves as the backbone for AI systems, enabling them to learn, adapt, and make informed decisions. However, the data utilized in these processes often come from heterogeneous sources, making it imperative to cleanse and prepare them before they can be effectively used.

The quality and integrity of data directly influence the performance of AI systems. Poor data quality can lead to inaccurate or biased AI models, which can have detrimental effects, especially in critical sectors like manufacturing. Inaccurate predictions or decisions based on flawed data can lead to production inefficiencies, increased costs, and even safety concerns. Therefore, ensuring the high quality of data is paramount for the successful implementation of AI in manufacturing processes.

The following steps can be utilized—

The first step involves gathering relevant data from various sources. This is challenging due to the volume of data, the diversity of sources, and the need for representative samples.

Once collected, data must be cleaned, transformed, and integrated. This step is crucial as it addresses missing values, outliers, inconsistencies, and other data quality issues.

Effective management ensures that AI systems can access and process data efficiently. This involves maintaining storage and retrieval systems, implementing version control, and ensuring data security and privacy.

This plays a crucial role in maintaining and enhancing data quality. It encompasses processes, policies, standards, and technologies that manage the availability, usability, integrity, and security of data. Effective data governance aids in decision-making, optimizes operations, ensures regulatory compliance, and offers a competitive edge.

As AI continues to transform the manufacturing sector, the importance of data quality becomes even more pronounced. By investing in robust data cleansing and preparation processes, manufacturers can ensure that their AI-driven processes are reliable, efficient, and poised to deliver optimal results.

5.1.4 Handling Missing Data, Anomalies, and Biases in Data

In AI-driven manufacturing processes, data is the backbone that drives decision-making, predictions, and optimizations. However, real-world data is often imperfect, containing missing values, anomalies, and biases. Addressing these imperfections is crucial for ensuring the accuracy and reliability of AI models.

In datasets, missing values refer to the absence of data for certain variables or observations. These can arise due to various reasons, such as equipment malfunctions, data entry errors, or participants not providing information. Missing values can skew statistical analyses, leading to biased or incorrect results. Anomalies, on the other hand, are unusual or unexpected data points that deviate significantly from other observations. Biases in data can arise from various sources, including sampling bias, measurement bias, and algorithmic bias, which can lead to unfair or discriminatory outcomes in AI models.

Unaddressed missing data can lead to inaccurate predictions, reduced model performance, and flawed decision-making. Anomalies, if not detected, can distort the overall understanding of the data, leading to misleading insights. Biases, especially in AI-driven manufacturing, can result in unfair treatment, misallocation of resources, and potential reputational damage.

Table 5.2 discusses the challenges of managing data in AI-driven manufacturing with details on how to handle missing values, anomalies, and biases.

5.2 Achieving Model Explainability and Accountability for Ethical AI

The global explainable AI market size was valued at USD 5.49 billion in 2022 and is expected to grow at a compound annual growth rate (CAGR) of 18.0% from 2023 to 2030. Explainable AI (XAI) refers to the development and deployment of artificial intelligence systems that can provide human-interpretable explanations for their decision-making processes. While traditional AI models like deep neural networks can achieve high accuracy in tasks such as image recognition or natural language processing, they often lack transparency and can be considered "black boxes" due to their complex internal workings [2].

With the increased adoption of AI-driven manufacturing processes, machine learning models are being used to optimize processes, predict outcomes, and automate tasks. However, as these models are deployed at scale across the entire end-to-end manufacturing process, there is a growing need to ensure they operate transparently and ethically. This section explores into the importance of model explainability and accountability, ensuring that AI not only enhances manufacturing processes but does so understandably and responsibly.

Table 5.2 Strategies for managing data challenges in AI-driven manufacturing—addressing missing values, anomalies, and biases with practical solutions

Topic	Description	Solutions and approaches
Dealing with missing values	• Absence of data for certain variables or observations due to reasons like equipment malfunctions or data entry errors • It can skew statistical analyses	1. Deleting rows with missing values 2. Imputation using mean, median, mode, or the most frequent value 3. Predictive modeling using non-missing values 4. Unsupervised techniques like K-Mcans or Hierarchical clustering
Detecting and addressing anomalies	• Unusual or unexpected data points that deviate significantly from other observations • It can distort the overall understanding of the data	1. Statistical methods for detection 2. Visualization techniques 3. Machine learning models for anomaly detection 4. Correcting, removing, or flagging anomalies once identified
Mitigating biases	• Arises from sources like sampling, measurement, and algorithmic processes • It can lead to unfair or discriminatory AI model outcomes	1. Understand the source of bias 2. Review data collection processes 3. Adjust sampling methods if necessary 4. Use fair and interpretable machine learning models 5. Regular audits and feedback loops for continuous monitoring

What is Model Explainability and Accountability?

Model explainability refers to the ability to understand and interpret the decisions made by AI models. In the world of AI, where complex algorithms often operate as "black boxes," techniques like LIME (Local Interpretable Model-agnostic Explanations) and SHAP (SHapley Additive exPlanations) have emerged as tools to demystify these models. They provide insights into how models make decisions, highlighting the significance of different input features.

Accountability, on the other hand, focuses on the ethical and responsible use of AI. It involves ensuring that models are developed, trained, and deployed with considerations for fairness, transparency, and without inherent biases. It's about taking responsibility for the model's actions and ensuring they align with ethical standards.

Why is it Crucial in AI-Driven Manufacturing?

Manufacturing processes are intricate, with numerous variables at play. Decisions made by AI can have profound implications, from the quality of the product to the safety of the production line. When stakeholders, be they engineers, managers, or floor workers, understand how decisions are derived, they're more likely to trust and integrate AI solutions into their workflows. Interpreting model decisions can highlight areas of improvement, both in the model itself and in the manufacturing process it influences.

With a clear understanding of model decisions, manufacturers can ensure that AI-driven processes are not only efficient but also ethically sound, avoiding potential biases or unfair practices. To achieve model explainability and accountability, manufacturers must adopt a multi-faceted approach encompassing technical, procedural, and ethical strategies as discussed below:

Tools like LIME and SHAP should be integrated into the AI development process. They can help data scientists and engineers understand model decisions, refine algorithms, and communicate findings to non-technical stakeholders. A thorough documentation process that captures the model's design, training data, methodologies, and performance metrics can serve as a reference point, ensuring transparency and facilitating audits or reviews.

Regular ethical reviews of AI models, considering their societal, economic, and environmental impact, can ensure that they align with broader ethical standards and values.

5.2.1 Interpretability Techniques Like LIME and SHAP

In AI-driven manufacturing processes, the deployment of machine learning models has become commonplace. But these deployed models are often complex and difficult to understand. As the manufacturing industry increasingly relies on AI to make critical decisions, the need for transparency and interpretability in these models becomes paramount. Model interpretability techniques like LIME (Local Interpretable Model-agnostic Explanations) and SHAP (SHapley Additive exPlanations).

5.2.1.1 What Are LIME and SHAP?

LIME and SHAP are advanced techniques designed to explain the predictions of machine learning models. While machine learning models, especially deep learning ones, are often termed "black boxes" due to their opaqueness, LIME, and SHAP aim to shed light on these boxes, making their decisions understandable to humans.

LIME focuses on explaining individual predictions. It perturbs the input data, observes the changes in predictions, and then approximates a simpler, interpretable model around the prediction to provide an explanation.

SHAP, on the other hand, is rooted in cooperative game theory. It assigns each feature an importance value for a particular prediction, ensuring consistent and fairly distributed feature attributions.

Interpretability in AI-driven manufacturing processes has become very important as manufacturing processes become more automated and data-driven, the decisions made by AI models directly impact production quality, efficiency, and safety.

5.2.2   Why Understanding AI Decision-Making Enhances Trust and Compliance?

Stakeholders, including engineers, technicians, and managers, are more likely to trust and adopt AI solutions if they understand how decisions are made. If a model's prediction is incorrect or suboptimal, understanding its reasoning can help in diagnosing and rectifying the issue. In regulated industries, being able to explain AI decisions can be crucial for compliance and auditing purposes.

5.2.2.1   How Are LIME and SHAP Implemented?

LIME works by approximating the complex model with a simpler, interpretable one for individual predictions. It does this by perturbing the input data, observing how predictions change, and then fitting a simpler model to explain those changes. The resulting explanation highlights the features that influenced the prediction the most.

SHAP values, derived from Shapley values in game theory, provide a unified measure of feature importance. By considering all possible feature combinations, SHAP values ensure a consistent and fair distribution of feature attributions for a given prediction.

5.2.2.2   Example Use Cases of LIME, SHAP in AI-Driven Manufacturing

An AI model might suggest altering production schedules to reduce energy consumption during peak hours. Using LIME, manufacturers can understand which factors (e.g., machine usage patterns, energy pricing) most influenced this decision, ensuring it's not only cost-effective but also environmentally responsible.

An AI system might flag a product as defective. With SHAP, manufacturers can pinpoint which features (e.g., color, shape, texture) were deemed most significant in this classification, ensuring that the system's decisions align with human expertise.

5.2.3 Documenting the Model Development Process

The development of machine learning models is a complex process. These models are created from a step-by-step process involving data collection, preprocessing, model training, validation, and deployment. As these models are being increasingly adopted for critical decision-making and process optimization in the manufacturing industry, it is imperative to maintain a comprehensive record of their development. A thorough documentation process can help collaborate and reproduce the model development process in the future.

Documentation refers to the systematic recording of all aspects related to the development of a machine-learning model. It encompasses everything from the initial problem statement, data sources, preprocessing techniques, model architecture, training parameters, and performance metrics to the final deployment details. Modern tools, such as Google's Document AI and Hugging Face's model cards, offer platforms to create, maintain, and share such Documentation, ensuring transparency and reproducibility.

Hugging Face has a vast collection of over 60,000 pre-trained models across various domains and languages, available on the Hugging Face Hub for flexible machine-learning applications [3].

5.2.3.1 Why is Documentation Crucial in the AI Process in Manufacturing?

Manufacturing processes are inherently complex, with numerous variables influencing outcomes. When AI models are introduced to optimize these processes, stakeholders need assurance regarding their reliability, efficacy, and safety. Documentation serves multiple purposes—

Transparency:

It provides a clear view of how the model was developed, the data it was trained on, and the logic behind its decisions.

Reproducibility:

With comprehensive Documentation, other data scientists or engineers can replicate the model, validate its results, or even improve upon it.

Accountability:

In case of discrepancies or failures, Documentation serves as a reference point to trace back and identify potential issues in the model development process.

5.2.3.2   How to Document the Model Development Process?

Comprehensive and detailed documentation is paramount to ensure transparency and reproducibility in AI model development. The key steps for documenting the model development process are discussed below:

Platforms like Google's Document AI offer automated solutions to extract, process, and categorize information from various sources, making the documentation process more efficient. Hugging Face's model cards provide a standardized format to document and share details about machine learning models, ensuring consistency and comprehensiveness.

Every aspect of data collection, including the sources, sampling methods, preprocessing techniques, and data augmentation strategies, should be meticulously documented and recorded, including the use of a data versioning tool. This ensures that the model's training data is transparent and can be replicated or audited if needed.

The choice of model architecture, hyperparameters, training algorithms, and validation techniques should be detailed. Tools like Modin with Scikit-learn can accelerate the model development process, and their usage should be documented. Metrics like accuracy, precision, recall, and any custom metrics used should be recorded along with details of the validation set and any cross-validation techniques employed.

Information about how the model is deployed, whether on-premises, in the cloud, or at the edge, should be included. Any post-deployment monitoring or updating strategies should also be documented.

5.2.3.3   Examples of Documentation

The documentation examples demonstrate how maintaining thorough records of model development, performance, and data sources enables vital explainability into adopted AI systems, establishing trust and responsible augmentation of manufacturing processes through transparent insights versus inscrutable black boxes.

Predictive Maintenance:

For a model predicting machinery failures, documentation would detail the sensor data sources, any preprocessing to remove noise, the choice of time series model used, training cycles, and performance on validation data. Tools like Document AI from Google could automatically extract relevant details from maintenance logs to enrich the documentation.

Quality Control:

In a model identifying defects using camera feeds, the documentation would describe the image data collection process, any data augmentation techniques, the convolutional neural network architecture used, and its performance metrics. Hugging Face's model cards could provide a standardized overview of the model's capabilities and limitations.

5.2.4 Communicating the Model's Capabilities and Limitations

In spite of the increased adoption of AI in manufacturing processes, these AI-based models have their own set of capabilities and limitations. Communicating these aspects is crucial for stakeholders to make informed decisions and set realistic expectations.

In manufacturing, AI models are capable of performing a wide range of tasks, from predictive maintenance and quality assurance to supply chain optimization and demand forecasting. These models can process vast amounts of data, identify patterns, and make predictions or decisions in real-time. They can adapt to changing conditions, learn from new data, and even automate complex processes.

However, AI models have their limitations. They require large amounts of data to train effectively, and the quality of this data directly impacts their performance. They can sometimes be seen as "black boxes," making it challenging to understand their decision-making processes. Moreover, AI models can be susceptible to biases, can't think critically or creatively like humans, and might not always account for rare or unforeseen events.

5.2.4.1 Why is Communicating These Aspects Crucial?

Understanding the capabilities of AI models allows manufacturers to utilize their full potential, optimizing processes, reducing costs, and enhancing product quality. On the other hand, being aware of their limitations ensures that manufacturers don't over-rely on these models, setting realistic expectations and implementing necessary safeguards.

Furthermore, clear communication fosters trust among stakeholders. When decision-makers, technicians, and workers understand what AI can and cannot do, they are more likely to accept and adopt AI-driven solutions. It also ensures that when things go wrong, teams can quickly identify whether the issue arose from the AI model's limitations or other external factors.

5.2.4.2 How Can Manufacturers Communicate These Aspects Effectively?

Effective communication strategies are essential to ensure staff understanding and trust in AI systems within an organization. Regular workshops and training sessions can be organized to educate staff about the AI models in use, their capabilities, and their limitations.

Detailed documentation that outlines the model's design, data sources, training methods, expected performance, and known limitations can be made readily available. Visual dashboards can be developed to provide real-time insights into the model's performance, highlighting its predictions, confidence levels, and any potential areas of concern.

5.2.4.3 Examples of Challenges in AI-Driven Manufacturing

These examples highlight that even as AI models deliver transformative capabilities, inherent limitations persist that teams must proactively identify and mitigate. Without explainability around model logic and clear communication of constraints to users, those leveraging AI for optimization risk over-reliance, undetected biases, and real-world performance gaps compared to promised potential.

In predictive maintenance and quality assurance alike, stakeholders knowing precisely where AI delivers advantages along with vulnerabilities enables balanced human and AI collaboration.

Predictive Maintenance:

While an AI model might accurately predict when a machine is likely to fail based on historical data and sensor readings, it might not account for a sudden external factor, like a power surge. Communicating this limitation ensures that technicians also rely on manual inspections and not solely on AI predictions.

Quality Control:

An AI-driven camera system can detect defects in products with high accuracy. However, if the system was not trained on a particular type of defect because it's rare, it might miss it. Ensuring workers are aware of this limitation can lead to additional manual checks for comprehensive quality assurance.

5.3 Overcoming Technical Gaps in Embedding AI Within Legacy Systems and Workflows

Manufacturing processes have traditionally been anchored on legacy systems, which, while reliable, are often outdated and not conducive to rapid technological advancements. Such systems, built using older technology, have been in use for decades and represent significant investments in terms of development, training, and process customization. As the manufacturing sector increasingly leans towards AI-driven processes for efficiency and innovation, there's a need to rejuvenate and transform these legacy systems without disrupting existing workflows.

Despite the clear advantages of transitioning from legacy systems to modern, AI-driven platforms, many enterprises, both in the public and private sectors, are hesitant. This hesitancy stems from the perceived risks of the transformation process, which includes the complexities of legacy systems, the potential for performance issues, the challenge of mapping current systems to new architectures, and the often-scant documentation of legacy systems. Given these challenges, there's an increasing emphasis on finding methods for safe and efficient legacy system transformation, ensuring that manufacturing processes can benefit from the latest in AI and data analytics without risking current operations.

To address the challenges associated with integrating AI into legacy systems, organizations can adopt a phased approach that minimizes disruption while ensuring measurable progress. The process can start with a comprehensive assessment of existing systems to identify integration points and determine the readiness for AI implementation. By mapping out critical dependencies and documenting system workflows, businesses can mitigate risks associated with transforming their technology infrastructure.

Moreover, a hybrid approach can be adopted, offering a flexible solution where legacy systems coexist with modern AI-driven components through middleware, APIs, and modular enhancements. This incremental strategy can allow manufacturers to run pilot tests on AI capabilities in isolated workflows before full-scale deployment, ensuring stability and compatibility with existing operations.

Organizations can also leverage partnerships with AI solution providers to tailor AI tools for specific manufacturing needs. These collaborations play a crucial role in supporting businesses, enabling them to bypass the steep learning curve associated with AI technologies and offering bespoke solutions that seamlessly fit into legacy workflows. Additionally, investing in workforce training ensures that operators and engineers can efficiently use AI tools, bridging the gap between traditional processes and cutting-edge technology.

The next sections discuss practical solution implementation strategies for facilitating this transition, focusing on processes/tools such as MLOps for end-to-end automation, API-based incremental integration strategies, and containerization for scalable deployment. These solutions can not only streamline the adoption of AI but also address challenges related to compliance, security, and operational continuity.

5.3.1 End-to-End Automated Machine Learning Workflows Using MLOps

Machine Learning Operations, commonly referred to as MLOps, is a set of best practices and tools that unify machine learning (ML) system development and operations (Ops). It's a discipline that bridges the gap between the development of machine learning models and their deployment into production. In AI-driven manufacturing processes, MLOps ensures that AI models are not just developed but also seamlessly integrated, monitored, and maintained within the manufacturing workflow.

Figure 5.2 shows the MLOps process includes three broad phases: Designing the ML-powered application, ML Experimentation and Development, and ML Operations.

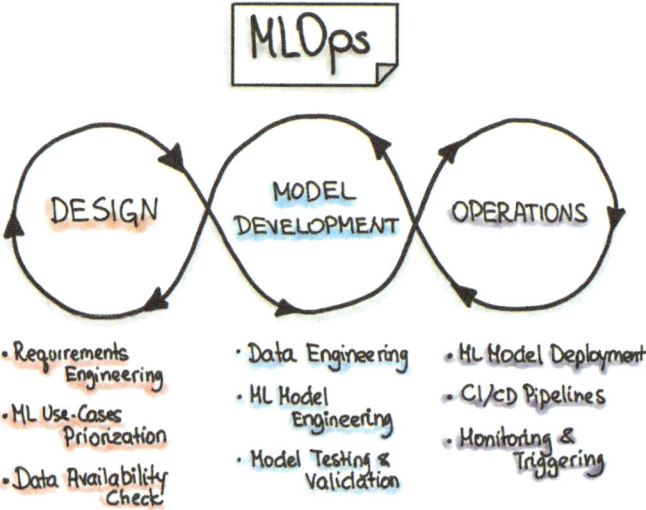

Fig. 5.2 The complete MLOps process [4]. *Source* https://ml-ops.org/content/mlops-principles. It is published under Creative Commons Attribution 4.0 International Public License and can therefore be shared and adapted with attribution ("INNOQ"). https://creativecommons.org/licenses/by/4.0/

For AI to truly transform manufacturing processes, these models must be effectively deployed, monitored, and iteratively improved upon in real-time operational environments. MLOps facilitates this by offering a structured framework for automating, deploying, and monitoring machine learning workflows. It ensures models are robust, up-to-date, and deliver consistent value.

Figure 5.3 shows the automated architecture for MLOps implementation CI/CD pipelines.

MLOps, especially when accelerated with external tools like Modin and Scikit-learn, provides an efficient framework for end-to-end automation of machine learning workflows.

5.3.1.1 Enhancing Manufacturing Efficiency with MLOps

Incorporating Machine Learning Operations (MLOps) into manufacturing enhances efficiency and accuracy in various stages of AI model deployment and management. The steps are discussed below:

With tools like Modin, data processing, which is often a time-consuming step in ML, becomes significantly faster. In manufacturing, where vast amounts of data are generated, this means quicker insights from data such as equipment logs, quality checks, or supply chain metrics.

Scikit-learn, a powerful tool for ML, when combined with the efficiencies of MLOps, ensures that models are trained, validated, and tested efficiently. For

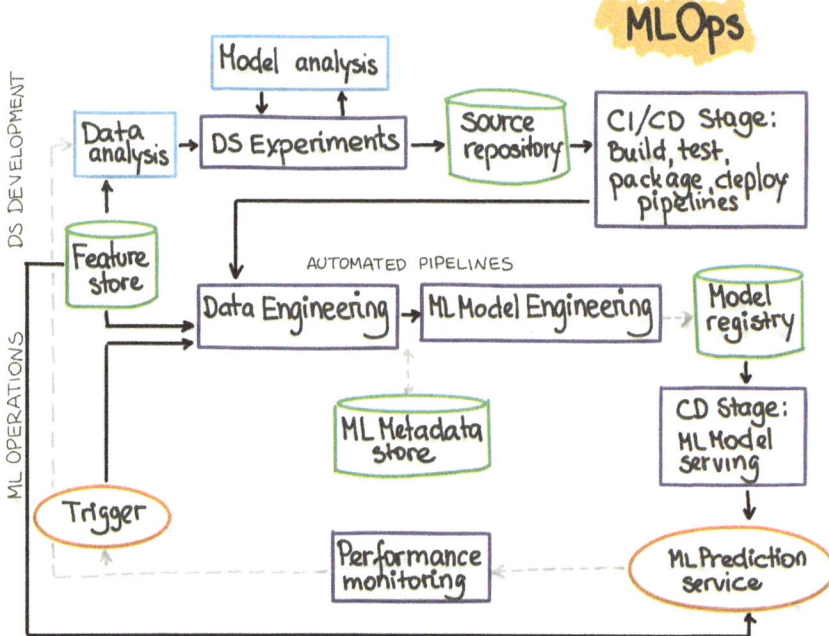

Fig. 5.3 Automated ML pipeline architecture with CI/CD routines [5]. *Source* https://ml-ops.org/content/mlops-principles. It is published under Creative Commons Attribution 4.0 International Public License and can therefore be shared and adapted with attribution ("INNOQ"). https://creativecommons.org/licenses/by/4.0/

instance, a model predicting machinery maintenance needs can be continually trained on new data, ensuring predictions are always accurate.

MLOps supports continuous integration and deployment, ensuring that once a model is trained, it's promptly deployed into the manufacturing workflow. This real-time integration means that insights derived from AI models can be acted upon immediately.

One of the cornerstones of MLOps is the continuous monitoring of deployed models. In manufacturing, this ensures that the AI tools being used are always performing optimally. If a model's performance starts degrading, perhaps due to changing conditions on the factory floor, MLOps frameworks can trigger retraining or alerting mechanisms.

5.3.1.2 Example Use Cases

In a manufacturing setup, equipment health is crucial. MLOps can automate the entire workflow of collecting machine data, predicting when a piece of equipment might fail, and then either automatically scheduling maintenance or alerting the necessary personnel.

Automated visual inspection systems can use AI models to detect defects in products. MLOps ensures that these models are always up-to-date, integrating new data into the training set and deploying improved models in real time.

5.3.2   Strategies for Incremental Integration like APIs

Application Programming Interfaces (APIs) serve as intermediaries that allow software applications to communicate with each other. In AI-driven manufacturing processes, APIs can play an important role in ensuring that various components, including legacy systems, modern applications, and AI models, can seamlessly interact, share data, and execute functions. An "API-first" strategy ensures that subsequent development aligns with the integration needs of the manufacturing process.

Figure 5.4 illustrates how an API communication pathway for deploying an AI application.

Manufacturing environments are diverse ecosystems of tools, machinery, software applications, and data streams. With the advent of AI, the need for these components to interoperate has become even more pressing. AI models, for instance, need data from sensors on the factory floor to predict maintenance needs or from inventory systems to optimize the supply chain. APIs provide a standardized way for these data exchanges to occur, ensuring that AI-driven insights are timely, accurate, and actionable.

Furthermore, leveraging APIs with tools like MuleSoft and integrating AI capabilities directly within the APIs can supercharge the manufacturing process, bringing in smart integrations and real-time intelligence.

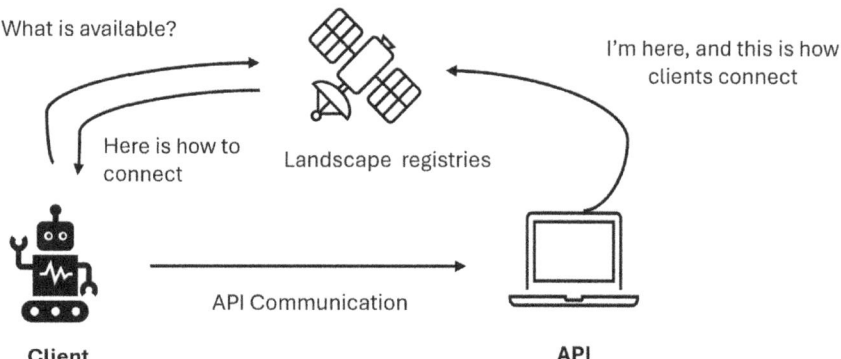

Fig. 5.4 Conceptual API communication for AI payload delivery. *Source* Created by the authors in Power Point

5.3.2.1 How to Implement APIs for AI Applications in Manufacturing

Designing with an "API-First" Strategy: Beginning with a clear API design ensures that the subsequent software development caters to the integration needs of the manufacturing environment. This not only ensures seamless data flow but also accelerates the development process.

Smart Integration with MuleSoft: MuleSoft offers a platform for designing, building, and managing APIs. In AI-driven manufacturing, MuleSoft can be used to integrate AI capabilities within the APIs, ensuring real-time data processing, analytics, and insights.

Embedding AI in APIs: AI capabilities can be directly embedded within the APIs. This means that as data flows through the API, AI models can process it in real time, offering immediate insights or actions. For instance, as data from a sensor on a machine flow through the API, an embedded AI model can immediately assess if the machine is likely to fail soon.

5.3.2.2 Example Use Cases

Cameras on the manufacturing floor capture images of products. These images are sent via an API to an AI model, which checks for defects. If a defect is detected, the information is immediately sent back through the API to flag the product.

Sensors on machinery send data via an API to an embedded AI model. The model, in real-time, predicts when a machine part might fail and sends back this prediction, allowing for preemptive maintenance.

5.3.3 Containerization and Microservices Architecture

Containerized microservices refer to the architectural approach where software applications are developed, packaged, and deployed as isolated, lightweight, and scalable containers. These containers are stand-alone units that encapsulate specific functionalities, ensuring that each microservice can function independently yet still communicate with others. In AI-driven manufacturing processes, containerized microservices provide a flexible and efficient method to integrate AI tools, applications, and data sources.

Figure 5.5 illustrates the conceptual architecture of deploying AI/ML applications in a containerized microservices architecture environment.

Manufacturing processes integrated with AI have multiple components, including software applications, AI models, data processing tools, and integration pipelines. Ensuring that each component can operate in harmony, scale on demand, and be updated or replaced without disrupting the entire system is crucial.

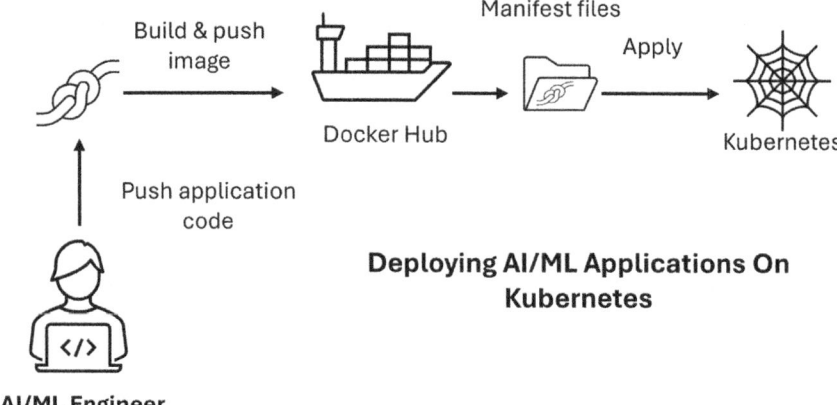

Fig. 5.5 Deploying AI application with container and microservice architecture. *Source* Created by the authors in Power Point

Containerized microservices offer several advantages. They provide a modular approach, ensuring that each container can be developed, deployed, scaled, or replaced independently. Moreover, the containerization of AI tools ensures they can be seamlessly integrated into the manufacturing DevOps pipeline, streamlining development, deployment, and monitoring.

5.3.3.1   How to Containerize AI Applications?

Containerization is a useful technique for deploying AI applications efficiently and effectively in diverse environments.

The steps are discussed below:

Containerized microservices offer a high degree of modularity. Each microservice, packaged in its container, handles a specific function. In AI-driven manufacturing, this could mean having separate containers for predictive maintenance, quality assurance, and supply chain optimization. Containers can be easily replicated or scaled based on demand. If a particular AI model experiences a surge in data input, its container can be scaled to handle the load without affecting other services.

The containerized AI tools can be seamlessly integrated into the DevOps pipeline. Containers provide a consistent environment, ensuring that AI models developed in a particular setting will operate the same way when deployed in the manufacturing process. Containers support rapid deployment and iterative development. If a new version of an AI tool is developed, it can be packaged into a new container and deployed without disrupting existing processes.

5.3.3.2 Example Use Cases

In a manufacturing setup, a containerized microservice can be dedicated to predicting product demand based on various factors. As market conditions change, this service can be scaled or updated independently.

Cameras and sensors on the manufacturing floor capture data. This data is sent to a containerized AI model, which analyzes it in real-time to detect defects or anomalies.

5.3.4 Maintaining Regulatory Compliance

Regulatory compliance refers to adhering to laws, regulations, guidelines, and specifications relevant to business processes. In AI-driven manufacturing processes, it means ensuring that the application and deployment of AI tools and models comply with established regulations, especially those concerning safety, quality, data privacy, and ethical use.

The integration of AI introduces new challenges related to ethics, data privacy, and potential biases. Manufacturers are becoming increasingly accountable for ensuring that their AI deployments do not violate regulatory standards. Non-compliance can lead to hefty penalties, reputational damage, and even operational setbacks. Beyond the legal implications, maintaining compliance ensures that AI tools enhance manufacturing processes without compromising on safety, ethics, and quality.

5.3.4.1 Solution Strategies for Ensuring Regulatory Compliance

To effectively maintain regulatory compliance in AI-driven manufacturing processes, manufacturers should consider adopting the following strategies:

As AI in manufacturing is relatively new, the regulatory landscape is continually evolving. Manufacturers must stay updated with the latest regulations, be it local, national, or international. This involves investing in legal expertise and regularly auditing AI tools for compliance. Ethical considerations should be at the forefront of AI integration. This involves ensuring that AI models are transparent, free from biases, and do not inadvertently introduce discriminatory practices in the manufacturing process.

AI models in manufacturing often rely on vast amounts of data. Manufacturers must ensure that data collection, storage, and processing adhere to privacy regulations. This includes securing data to prevent breaches and being transparent about data usage. Given the dynamic nature of AI and the evolving regulatory landscape, manufacturers should establish a system for continuous monitoring and auditing of their AI tools. This ensures that any deviations from compliance can be promptly detected and rectified.

Employees should be trained and made aware of the importance of regulatory compliance in AI-driven manufacturing. This fosters a culture of compliance and ensures that everyone understands their role in maintaining it.

5.4   Developing Internal Skills and Changing Cultures

The integration of AI within the information technology (IT) workforce is progressing at a rapid pace. A report by Bergur Thormundsson, shows that in 2022, 47% of IT executives anticipated widespread AI adoption within their companies. Furthermore, 20% of IT professionals anticipated that AI is critical, and this number is expected to grow to 49% by 2025 [5].

An AI-driven manufacturing organization needs more than just integrating advanced technologies, it requires a holistic approach that encompasses upskilling employees, nurturing a culture of continuous learning, and fostering an environment that embraces change and innovation.

In order to realize the potential of AI, organizations need to go beyond technology. Employees need the skills to utilize AI's power, while organizational cultures need to be agile, promoting innovation and adaptability. Focusing on these aspects ensures that the organization not only adopts AI but thrives with it, realizing tangible improvements in efficiency, productivity, and innovation.

5.4.1   Vision for Upskilling Employees to Enhance Manufacturing Through AI

To truly leverage the power of AI in manufacturing processes and ensure sustainable transformation, organizations should consider the following strategies:

Organizations should prioritize skills-based training. This involves identifying crucial skills for the future, offering training programs, and ensuring employees are equipped to navigate an AI-driven environment. It is important to foster a culture that encourages experimentation. In an AI-driven manufacturing setup, this might involve setting up innovation labs, promoting cross-functional collaborations, or even encouraging employees to work on side projects.

AI tools can identify skill gaps, recommend personalized learning pathways, or even predict future skill requirements, ensuring the workforce remains ahead of the curve.

5.4.2 Solution Implementation Strategies

Several key strategies for implementing solutions to achieve the vision are discussed below:

Personalized Learning Paths:

Using AI, a manufacturing organization can analyze an employee's current skill set, role requirements, and personal interests to recommend a tailored learning path. This ensures relevant skill development and keeps employees engaged.

Predictive Skill Requirement:

AI tools can analyze industry trends, internal project requirements, and technological advancements to predict future skill needs. This allows organizations to train their workforce proactively.

Cross-functional Collaboration Platforms:

AI-driven platforms can identify potential collaborations between departments, promoting knowledge sharing and holistic project development.

5.4.3 Change Management Strategies for AI Adoption

Change management for AI adoption involves a structured approach to transitioning individuals, teams, and organizations from their current state to a desired future state where AI seamlessly integrates with and augments manufacturing processes.

Figure 5.6 illustrates the impact of planned change management from the adoption of AI.

In manufacturing, the integration of AI tools and methodologies can also be met with resistance, apprehension, or misunderstanding by employees. The success of AI adoption in manufacturing isn't just about the technology itself but also about how people—ranging from the shop floor to the boardroom understand, embrace, and utilize it. Effective change management ensures that the organization navigates this transition smoothly, with all stakeholders aligned and equipped to utilize the benefits of AI.

5.4.3.1 Change Management Strategies for Successful AI Adoption

To drive successful AI adoption in manufacturing processes, organizations should prioritize the following change management strategies.

It's crucial to involve all relevant stakeholders, from leadership to frontline workers, in the AI adoption journey. Their input, concerns, and feedback can provide invaluable insights, ensuring the AI implementation is relevant and user-friendly. It is

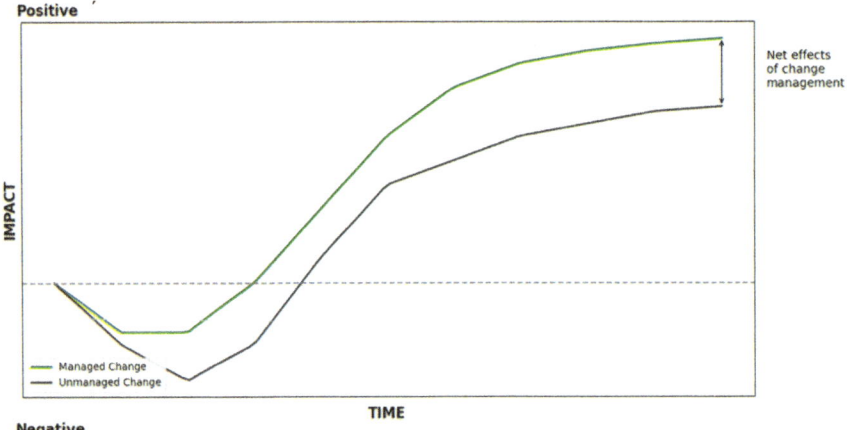

Fig. 5.6 Illustrative guide showing the impact of change management when implemented. *Source* Created by the authors in Python Notebook

important to have regular and transparent communication about what AI is, why it's being adopted, and how it will impact existing processes. This helps dispel myths, address concerns, and build confidence in the AI-driven approach.

Introducing AI in manufacturing requires equipping employees with new skills and knowledge. Tailored training sessions, workshops, and seminars can help the workforce understand and effectively use AI tools. Implementing mechanisms to gather feedback allows organizations to understand pain points, challenges, and areas of improvement. This iterative approach ensures that the AI adoption process is continually refined based on real-world insights.

Recognizing and celebrating milestones, whether big or small, can boost morale and encourage further engagement with the AI transition.

5.4.3.2   Example Scenarios of Change Management Adoption

A manufacturing company introduces AI for real-time quality checks. Change management strategies involve workshops for quality control teams, feedback sessions after the first month of implementation, and celebrating a significant reduction in defects detected.

A facility introduces AI-driven predictive maintenance for its machinery. The change management process includes training sessions for maintenance teams, regular updates on machinery health to the entire plant, and recognizing a month with zero unplanned downtimes.

5.4.4 Promoting Cross-Functional Collaboration

Cross-functional collaboration refers to the active cooperation and teamwork between different departments or functional areas within an organization. In AI-driven manufacturing, it means ensuring that various teams—from data scientists to production engineers to marketing professionals—work together to leverage AI's capabilities fully.

Manufacturing processes are intricate, often involving multiple departments. As AI integration promises to revolutionize these processes, it's imperative that these departments do not continue to work in silos. Integrating AI isn't just a technological shift but also an organizational one.

In order to fully realize the benefits of AI, there needs to be a holistic understanding and application of the technology across all departments. This can only be achieved when there is active collaboration between different functions, ensuring that AI tools are developed, deployed, and utilized in a manner that addresses the specific needs and challenges of each department while also contributing to the organization's overarching objectives.

5.4.4.1 Solution Strategies for Enhancing Cross-Functional Collaboration

To foster effective cross-functional collaboration in AI-driven manufacturing processes, organizations might consider the following avenues:

Forming dedicated teams comprising members from different departments is crucial. Such teams can ensure that AI projects address diverse perspectives and needs, leading to holistic solutions. Workshops can be organized where teams share updates, discuss challenges, and brainstorm solutions, ensuring everyone is aligned and informed.

Setting clear, organization-wide objectives for AI adoption ensures that all departments are working towards a common goal. This can prevent conflicts and promote a more cooperative approach. Establishing channels where teams can provide feedback on AI tools and implementations ensures that the tools are continuously refined and improved based on real-world insights.

Creating platforms where teams can share knowledge, best practices, and learning can foster a culture of continuous learning and collaboration.

Example of cross-functional collaboration:

The maintenance, production, and data science teams work together to develop an AI-driven predictive maintenance system. A unified vision ensures the system not only predicts machinery breakdowns but also aligns with production schedules, minimizing disruptions.

Quality assurance, manufacturing, and IT departments collaborate to deploy AI for real-time quality checks. Regular brainstorming sessions lead to the development of a system that not only detects defects but also integrates with the existing IT infrastructure seamlessly.

5.4.5 Leadership Buy-in for AI Transformation

Leadership buy-in refers to the commitment, support, and active involvement of top-level management in the adoption and integration of AI into manufacturing processes. The adoption of AI in manufacturing is not just about introducing new technology; it represents a significant shift in how processes are approached, executed, and optimized. Successful AI integration is contingent upon the active support and involvement of the leadership team.

Leadership buy-in ensures that the AI transformation aligns with the organization's broader strategic objectives, garners the necessary resources, and navigates the inherent challenges and roadblocks effectively. Furthermore, when leadership champions AI initiatives, it sends a strong message throughout the organization about the importance and priority of the transformation, facilitating smoother adoption and integration at all levels.

5.4.5.1 How to Ensure Leadership Buy-in for AI Projects?

To secure leadership buy-in for AI transformation in manufacturing processes and ensure their continuous support throughout the journey, consider the following approaches:

One of the first steps is ensuring that the leadership understands what AI is, its potential benefits, and its implications. Tailored workshops, seminars, or training sessions can be instrumental in this regard. AI initiatives should not be standalone projects. They should be intricately linked with the organization's broader strategic goals. Demonstrating this alignment can help in securing leadership support.

Implementing pilot projects and showcasing their successes can provide tangible evidence of AI's potential benefits, making it easier for leadership to support larger, more comprehensive initiatives. It is important to have regular updates and feedback loops with leadership. This ensures they are always in the loop, can provide guidance when needed, and can make informed decisions regarding resource allocation and strategic direction. Sometimes, external experts or consultants can provide an unbiased perspective, validate the AI strategy, and help in securing leadership buy-in.

5.5 Key Takeaways

- Embedding AI capabilities directly within APIs allows for real-time data analysis and decision-making. This setup enables immediate operational responses, such as predictive maintenance alerts and defect detection notifications.
- Containerized microservices ensure that each component of the AI system can be independently scaled and maintained without affecting the overall system. This architecture facilitates the integration of AI tools into existing manufacturing processes, enhancing agility in deployments and updates.
- It's crucial to ensure that AI implementations comply with existing regulations concerning data privacy, safety, and ethics. Establish systems for ongoing compliance checks to adapt to evolving legal standards and maintain operational integrity.
- Invest in training programs to equip employees with the necessary skills to work effectively with AI technologies. Promote a culture of innovation and adaptability to support the integration and acceptance of AI within the organization.
- Encourage collaboration across different departments to ensure that AI solutions are developed and implemented in a manner that addresses varied operational needs. Align AI initiatives with the organization's strategic goals to enhance buy-in and effectiveness.
- Demonstrate how AI projects align with broader business objectives to secure support from top management. Leadership advocacy can drive organizational commitment and resource allocation for AI projects.
- Engage all stakeholders in the AI transformation process to gather insights and address concerns, enhancing overall acceptance and integration. Maintain open lines of communication about AI initiatives, their impacts, and benefits to ensure clarity and build trust.
- Keep detailed records of AI model development, performance, and data usage to ensure transparency and facilitate audits. Prioritize ethical considerations in AI development to prevent biases and ensure fairness in automated decisions.

Glossary

API (Application Programming Interface) A set of protocols and tools for building software applications, specifying how software components should interact.

Containerization A lightweight form of virtualization that involves encapsulating an application and its dependencies into a container that can run on any computing environment.

DevOps A set of practices that combines software development (Dev) and IT operations (Ops) to shorten the systems development life cycle while delivering features, fixes, and updates frequently in close alignment with business objectives.

Document AI A technology that uses artificial intelligence to analyze and extract information from documents.

LIME (Local Interpretable Model-agnostic Explanations) A technique that explains the predictions of any machine learning classifier in an interpretable and faithful manner.

Microservices An architectural style that structures an application as a collection of small autonomous services, modeled around a business domain.

MLOps (Machine Learning Operations) The practice of collaborating and communicating between data scientists and operations professionals to help manage the production ML lifecycle.

Modin A DataFrame library that scales data science workflows by changing a single line of code.

Service-Oriented Architecture (SOA) A style of software design where services are provided to other components by application components, through a communication protocol over a network.

SHAP (SHapley Additive exPlanations) A game theoretic approach to explain the output of any machine learning model.

References

1. Fortune Business Insights. (2023). Big Data analytics market size | Global Statistics Report [2028] [Internet]. www.fortunebusinessinsights.com. . Access date November 21, 2024. https://www.fortunebusinessinsights.com/big-data-analytics-market-106179
2. Explainable AI market size, share & growth report, 2030 [Internet]. www.grandviewresearch.com. 2022. Access date November 21, 2024. https://www.grandviewresearch.com/industry-analysis/explainable-ai-market-report
3. Model Hub docs [Internet]. huggingface.co. Access date November 21, 2024. https://huggingface.co/docs
4. ml-ops.org [Internet]. ml-ops.org. Access date November 21, 2024. https://ml-ops.org/content/mlops-principles
5. Global AI adoption rates within IT business 2022 [Internet]. *Statista.* 2022. Access date November 21, 2024. https://www.statista.com/statistics/1346631/global-ai-function-adoption-rates-business-it/

Chapter 6
AI Safety and Ethical Considerations

Abstract This chapter discusses the critical aspects of AI safety and ethical considerations in manufacturing. As AI continues to enhance efficiency and drive innovation, ethical deployment has become indispensable for ensuring equitable benefits, fairness, and trust. Industrial manufacturers are leading with 89% reporting the adoption of AI ethics policies as of 2021. This chapter examines key ethical challenges, including algorithmic biases, cybersecurity threats, job displacement, and compliance with global regulations. The chapter outlines strategies for ethical AI implementation, by ensuring transparency, interpretability, human oversight, and continuous monitoring. Real-world case studies demonstrate the complexities of ethical dilemmas and their resolutions. Furthermore, the chapter explores the governance framework needed for responsible AI, built on the pillars of people, processes, and technology. Regulatory variations across regions, such as the EU's AI Act and China's stringent compliance requirements, underscore the necessity of aligning AI systems with diverse legal landscapes. By prioritizing ethical practices, organizations can safeguard stakeholder trust, address potential risks, and ensure sustainable AI-driven transformations. The chapter concludes with actionable insights for fostering responsible AI use, from designing fairness-aware algorithms to aligning corporate policies with international standards, ensuring that technological advancements contribute positively to society.

Keywords Ethical AI practices · AI governance framework · Algorithmic bias mitigation · Human oversight in AI · Responsible AI deployment · AI safety compliance

The adoption of ethics policies for AI in various industries has become a crucial aspect of AI integration. Bergur Thormundsson's report shows the implementation rates of these ethics policies across different sectors in 2021. Industrial manufacturers emerged as frontrunners, with 89 percent of respondents from this industry reporting the existence of AI ethics policies within their organizations [1]. This trend reflects a proactive approach to addressing the ethical implications of AI technology.

Other industries, such as retail and technology, are also swiftly moving towards implementing ethical guidelines, illustrating a broader recognition of the importance of responsible AI use across diverse sectors.

The growing adoption of AI in manufacturing has led to improved efficiency and spurred innovation. However, with this transformation comes a responsibility to ensure that AI is used ethically and safely. This chapter explores the ethical considerations and safety measures that manufacturers must prioritize as they integrate AI into their operations.

AI in manufacturing is not just about algorithms and data; it's about ensuring that these technologies are used in ways that are beneficial and fair and do not inadvertently harm individuals or society. Ethical AI usage in manufacturing encompasses a broad spectrum of considerations, from ensuring algorithmic fairness to addressing concerns about job displacement and cybersecurity.

The ethical deployment of AI in manufacturing is crucial for several reasons. It ensures that the benefits of AI are equitably distributed and do not perpetuate or exacerbate existing inequalities. It safeguards against potential harms, such as biased decision-making or security breaches. Ethical AI practices foster trust among stakeholders, including employees, customers, and the broader public.

6.1 Solutions Strategies for Ethical AI Usage in Manufacturing

There are several key strategies that manufacturers can adopt to ensure ethical AI usage in manufacturing. First, ensuring Transparency and Interpretability involves making AI processes clear and comprehensible, allowing stakeholders to understand how decisions are made. This transparency is crucial in building trust and ensuring accountability. Second, Human Oversight is vital to ensure AI does not function autonomously; human intervention is necessary to identify and rectify errors or biases in AI systems.

Finally, Continuous Monitoring is essential for maintaining ethical standards. This involves regular audits and updates to AI systems, ensuring they remain unbiased and effective over time. Together, these strategies form a comprehensive approach to ethical AI usage in manufacturing, balancing technological innovation with responsibility and human-centric values.

6.1.1 Case Studies on Resolving Ethical Dilemmas

This section discusses various case studies for a deeper look into ethical dilemmas in AI implementations in manufacturing and provides insightful examples of how complex ethical challenges are addressed [2].

Bias and Fairness

In the manufacturing industry, AI-driven decisions can inadvertently perpetuate biases present in the training data, leading to discriminatory outcomes. This is particularly alarming when decisions influence critical aspects like product quality or employment opportunities. Addressing these issues involves implementing fairness-aware algorithms and conducting regular audits to ensure outcomes are free from unfair discrimination.

Transparency and Accountability

The inherent opacity of AI's decision-making processes, often referred to as "black-box" algorithms, poses significant ethical challenges. It is crucial for manufacturers to enhance the transparency and explainability of these systems. This involves making the decision-making processes of AI understandable and accessible, thereby facilitating accountability and engendering trust among stakeholders.

Job Displacement and Worker Reskilling

The introduction of AI in manufacturing may lead to job displacement due to automation. Ethical considerations demand thoughtful approaches to minimize adverse effects on workers. This includes investing in upskilling and reskilling programs to help employees transition to new roles that align with the evolving technological landscape.

Data Privacy and Security

The vast amounts of sensitive data processed by AI systems in manufacturing raise serious concerns about data privacy and security. To protect against unauthorized access and breaches, manufacturers must implement stringent data protection measures, such as robust cybersecurity protocols and compliance with data privacy regulations.

Regulatory and Compliance Challenges

As AI technologies become integral to manufacturing, aligning with regulatory standards is essential. This includes adhering to international guidelines such as the GDPR for data protection and developing internal compliance frameworks that ensure AI applications are both ethical and legal [3].

These case studies highlight the necessity of integrating ethical considerations throughout the lifecycle of AI systems in manufacturing. They provide practical insights into navigating the ethical complexities of AI deployment, ensuring that technological advancements benefit all stakeholders equitably and sustainably. For further reading, readers can look at Princeton Dialogues on AI and Ethics Case Studies published by Princeton University [4].

6.2 Addressing Bias, Cybersecurity, and Job-Related Risks

As organizations continue to embrace AI, the perception and prioritization of associated risks have evolved. Bergur Thormundsson's report sheds light on the changing landscape between 2019 and 2022. In 2022, cybersecurity emerged as the predominant concern for businesses implementing AI, reflecting the nascent nature of AI technologies and the need to safeguard proprietary information [5].

The integration of AI into manufacturing processes has significantly improved efficiency, scalability, and accelerated innovation. However, these benefits are accompanied by a set of challenges and risks. From algorithmic biases that can skew decision-making to cybersecurity vulnerabilities that can compromise sensitive data and the looming threat of job losses due to automation, the path to AI integration is fraught with hurdles. This section dives deep into these challenges and offers insights into how they can be effectively mitigated.

AI's potential in manufacturing is vast, from enhancing natural language processing capabilities to making better-informed decisions by processing vast amounts of data. Yet, the rapid adoption of AI also brings forth challenges. The workforce, especially those in high-skill jobs, faces the risk of job automation or significant changes in job design. Additionally, the inherent "black box" nature of AI systems can lead to unintentional violations of laws related to bias, fraud, or antitrust, posing legal and financial risks to companies.

Addressing these challenges is imperative to ensure that the benefits of AI are realized without compromising on fairness, security, and workforce stability. Ethical AI practices not only safeguard against potential harm but also foster trust among stakeholders, ensuring that AI-driven transformations are sustainable and beneficial in the long run.

6.2.1 Solutions Strategies for Ethical AI Adoption in the Workforce

In addressing the challenges brought about by AI adoption in manufacturing, there are several key strategies that can be effectively implemented to ensure a harmonious and ethical integration of technology in the workplace.

As AI reshapes the job landscape, there's a pressing need to invest in training programs that help workers transition to new roles. For instance, AI can be used to match workers' skills with suitable job openings within the same company, minimizing disruptions. However, the changing nature of employment contracts necessitates policies that promote skill acquisition, especially for workers at risk of automation.

While AI's potential to automate tasks is evident, there's a need to balance automation with augmentation. Public funds can be channeled to stimulate AI research that augments rather than automates work. This ensures that AI complements human capabilities, leading to enhanced productivity without sidelining the workforce.

Ensuring that AI systems are transparent and fair is crucial. Regulatory agencies must be equipped to oversee AI-driven workplaces and ensure compliance with national and international laws. This includes setting robust standards for algorithmic audits, building technical expertise within agencies, and crafting policies that are mindful of the challenges posed by AI.

6.2.2 Example Use Cases

A manufacturing firm uses AI for hiring. While the algorithm streamlines the recruitment process, regular audits are conducted to ensure that the system does not have inherent biases.

Another company employs AI for warehouse management. To ensure transparency, workers are provided insights into how the AI system operates, fostering trust and collaboration.

6.3 Implementing Ethical AI with Transparency and Oversight

The approach to implementing responsible AI in U.S. organizations has seen a distinctive trend in 2022, as documented by Bergur Thormundsson in their report. They mention that organizations adopting a holistic strategy are addressing AI responsibility concerns more effectively than those without such an approach [6].

As AI systems become more intricate and intertwined with manufacturing operations, there is a growing emphasis on implementing responsible AI practices. These practices ensure that AI enhances productivity and respects ethical standards, ensuring transparency, interpretability, and human oversight.

Responsible AI practices refer to a set of guidelines and methodologies that prioritize ethical considerations in the design, development, and deployment of AI systems. These practices encompass various dimensions, including ensuring that AI models are transparent in their operations, interpretable in their outcomes, and have human oversight to prevent unintended consequences.

The adoption of AI in manufacturing processes, while promising, can inadvertently lead to biases, inaccuracies, and unintended outcomes. For instance, an AI system that predicts machinery maintenance might inadvertently prioritize certain machines over others due to inherent biases in the training data. Such biases can lead to costly downtimes or even safety hazards. Hence, ensuring that AI systems are transparent and interpretable becomes crucial. Moreover, having human oversight ensures that there's a safety net, a human touch, to validate and verify AI decisions, especially in critical scenarios.

6.4 Responsible and Ethical AI Practices

Implementing responsible and ethical AI practices in manufacturing, such as ensuring transparency, interpretability, and human oversight, is vital for building trust and efficacy in AI systems, from energy optimization in manufacturing units to quality control in automotive production.

6.4.1 Transparency

Transparency in AI refers to the clarity with which an AI system operates. It's about ensuring that stakeholders understand how the AI system makes decisions. In manufacturing, this could mean providing clear documentation about how an AI-driven quality control system identifies defects or how an AI-based inventory management system predicts stock requirements.

A manufacturing unit uses an AI system to optimize energy consumption. The system's decisions, algorithms, and data sources are clearly documented and shared with the facility managers, ensuring they understand and trust the system's recommendations.

6.4.2 Interpretability

Interpretability is about ensuring that humans can understand the outcomes of an AI system. It's not enough for an AI system to provide a decision; it should also provide a rationale. For instance, if an AI system recommends changing a production schedule, it should also provide reasons, like anticipated demand spikes or predicted machinery downtimes.

A car manufacturing plant uses AI to inspect paint quality. When the system detects a defect, it not only flags it but also provides a visual representation highlighting the defect's nature and possible causes, making it easier for human inspectors to understand and rectify.

6.4.3 Human Oversight

No matter how advanced an AI system is, there should always be a provision for human intervention. This ensures that there's a check and balance mechanism, especially in scenarios where the AI's decision might have significant consequences.

In a pharmaceutical manufacturing unit, an AI system suggests adjustments to the drug formulation based on various parameters. However, every suggestion is reviewed and validated by a team of human experts before implementation, ensuring safety and efficacy.

6.5 Establishing Governance for Responsible AI Applications

AI has transitioned from being a business tool for experimentation to an integral component of enterprise strategy across various sectors, including manufacturing. The potential of AI to transform data into actionable insights, enhance decision-making, and amplify human capabilities is immense. Ensuring responsible, transparent, and explainable AI is paramount, especially as the outcomes of AI models become increasingly business-critical.

AI governance refers to the overarching process of directing, managing, and monitoring an organization's AI activities. It encompasses the establishment of a framework that ensures AI models and insights operate reliably, with clear visibility and accountability throughout their lifecycle. This governance framework aims to address the challenges of scaling AI, managing associated risks, and meeting regulatory and compliance obligations.

The global AI market is predicted to witness significant growth, potentially reaching over $500 billion by 2024. As AI's influence expands across sectors like manufacturing, the need for robust governance structures becomes even more critical. Without proper governance, organizations risk deploying "black box models" where the decision-making process is opaque, leading to potential biases, inefficiencies, and unperceived risks. Moreover, non-compliance with emerging AI regulations can result in hefty fines and reputational damage.

6.5.1 Key Pillars of AI Governance in Manufacturing

The key pillars of AI governance in manufacturing—people, process, and technology are crucial in shaping effective, responsible, and compliant AI applications, from predictive maintenance systems to quality control processes.

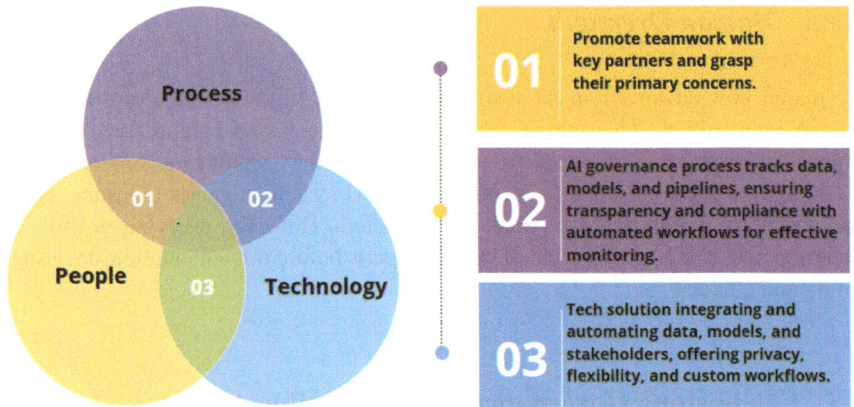

Fig. 6.1 Three pillars of AI governance in manufacturing. *Source* Created by the authors in Canva

Figure 6.1 shows the three pillars of AI governance in manufacturing, encompassing people, processes, and technology, forming the foundation for effective and responsible AI applications.

6.5.1.1 People

A successful AI governance framework requires a cross-functional team comprising various stakeholders, from data scientists to business leaders. Aligning these stakeholders is the first step, ensuring they understand the AI lifecycle, participate in ideation, and adopt responsible AI practices.

In a manufacturing company, the Chief Financial Officer (CFO), Chief Marketing Officer (CMO), Chief Risk Officer (CRO), and other key stakeholders collaborate to define performance metrics for AI-driven predictive maintenance systems, ensuring alignment with business controls and regulations.

6.5.1.2 Process

AI governance mandates comprehensive documentation of the AI model's lifecycle, from data origins to model deployment. This includes detailing the techniques used for training, hyperparameters, and testing metrics. Such documentation enhances transparency and helps organizations manage risks effectively.

A manufacturing plant uses AI for quality control. The AI governance process ensures that every decision made by the AI system, from identifying defects to recommending corrective actions, is traceable, documented, and auditable.

6.5.1.3 Technology

Implementing AI governance requires specific technological building blocks. An ideal solution should govern the end-to-end AI lifecycle, integrate diverse data sources, offer self-service access with privacy controls, automate model building and deployment, and provide support for customized governance workflows.

A manufacturing company adopts the IBM watsonx governance solution to manage its AI-driven supply chain optimization system. The solution offers comprehensive AI governance, from design to deployment, ensuring transparency, compliance, and risk management.

6.6 Ensuring Safety and Compliance with Regulators

The AI Compliance Monitoring Market is poised for rapid expansion in the United States. A report by Virtue Market Research, provides a detailed analysis of this market's growth. In 2022, the market was valued at USD 129.56 million, and it's projected to soar to a remarkable USD 1429.36 million by 2030. This growth, at an impressive CAGR of 35% during the forecast period of 2023–2030, can be attributed to the increasing demands for regulatory compliance and data privacy in various sectors [7].

This trend reflects the growing importance of AI in monitoring and ensuring compliance within businesses, a critical aspect in today's digitally driven and regulation-intensive business environment.

AI has become an integral part of the corporate world, offering numerous advantages such as automating routine tasks, optimizing decision-making, and enhancing efficiencies. In corporate regulatory compliance, AI-powered solutions are revolutionizing the way compliance officers operate. These tools can automate tasks, pinpoint risks, and analyze potential misconduct. However, the rapid evolution of AI also brings forth challenges that need to be addressed, especially when deploying these solutions in sensitive areas like manufacturing.

The rise of AI tools has not gone unnoticed by regulators. They are increasingly expecting, and in some cases demanding, that organizations utilize AI and data analytics to ensure compliance. For instance, the Department of Justice's (DOJ) guidelines on the Evaluation of Corporate Compliance Programs (ECCP) demonstrate the importance of leveraging corporate data to ensure the efficacy of a compliance program. The guidelines stress the importance of the compliance function having access to relevant data sources and utilizing this data to comprehend the risks a company might face.

Moreover, the public's concern about digital technology has shifted from the potential misuse of personal data to the actions of the software itself. Misapplied and unregulated AI can lead to unfair outcomes, primarily because it can magnify biases present in the data. This is especially pertinent in manufacturing processes where AI-driven decisions can have tangible impacts on product quality, safety, and compliance.

6.6.1 Leverage AI for Enhanced Compliance in Manufacturing

In manufacturing, leveraging AI for compliance involves identifying risks, monitoring programs, automating tasks, ensuring fairness, understanding AI logic, and adhering to regulations, thereby revolutionizing how compliance is managed in the industry.

The IEEE's 'Ethically Aligned Design: A Vision for Prioritizing Human Well-being with Autonomous and Intelligent Systems' (v1; 2019) is a significant document addressing ethical issues in AI and proposing mitigation strategies.

Identifying Potential Risks

AI can analyze vast amounts of data, such as transaction data and employee records, to pinpoint risks that might elude human analysts. For instance, in manufacturing, AI can detect anomalies in production data that might indicate quality issues or compliance breaches.

Monitoring Compliance Programs

AI can oversee compliance programs to ensure their effective implementation. It can review training records to ascertain that employees are well-versed with the company's compliance policies.

Automating Compliance Tasks

Mundane compliance tasks, like document reviews, can be automated using AI, allowing compliance officers to concentrate on strategic tasks. In manufacturing, AI can review quality assurance reports to detect potential non-compliance.

Ensuring Fairness and Addressing Bias

One of the primary challenges is the potential bias in AI models. If the training data is skewed, the model's decisions might also be biased. It's crucial to assess the data the model is trained on and evaluate the model's results for any signs of discrimination.

Understanding AI Logic

AI models can be intricate, making it challenging to decipher their decision-making process. Compliance officers need to understand the logic behind AI models to trust their results and explain their decisions.

Compliance with Regulations:

AI solutions must adhere to all relevant regulations, including data privacy laws. This is particularly challenging given the evolving nature of AI regulations across different jurisdictions.

Example Use Cases:

AI-powered risk scoring systems that review relationships between employees, vendors, and other third parties can help organizations identify behaviors that might violate regulations. For instance, in manufacturing, such systems can detect potential conflicts of interest in vendor relationships. Healthcare organizations can utilize AI to sift through clinical data, such as medication dispensed and treatments prescribed, to identify patients at increased risks for specific complications.

AI systems trained on the performance data of previous employees can screen job applicants. However, without proper oversight, these tools might inadvertently perpetuate biases against certain groups. As AI continues to permeate the corporate world, especially in manufacturing processes, it's essential for companies to proactively engage with regulators and ensure that their AI-driven processes are safe, compliant, and beneficial for all stakeholders.

6.6.2 Global Regulatory Variations and Impact on AI Implementation

The regulatory landscape for AI varies significantly across the globe, reflecting diverse approaches to balancing innovation with ethical and safety considerations. These variations can profoundly impact the development and deployment of AI technologies in different regions, including the United States, the European Union, and Canada [8, 9]. Regulations in China are tightly controlled by the Government [10].

United States

In the U.S., the approach to AI regulation is characterized by its sector-specific nature and a general hesitance to overregulate, which could potentially stifle innovation. The U.S. strategy focuses on advancing AI as a national economic resource, with federal initiatives aiming to promote trustworthy and ethical AI development across various departments. However, there has also been a move towards addressing algorithmic harms at the state level, with states like California and Vermont introducing their own legislation to tackle these issues.

European Union

The EU has taken a more structured approach with the AI Act, which aims to establish risk-based rules for AI usage. This act categorizes AI systems according to their risk levels, imposing stringent requirements on high-risk applications to ensure safety and compliance. The legislation emphasizes transparency and accountability, particularly concerning data privacy under the General Data Protection Regulation (GDPR).

United Kingdom

The UK's approach is proactive yet measured, focusing on using existing legislation and regulatory bodies to manage AI risks. This strategy aims to maintain flexibility and foster innovation by adapting existing legal frameworks to the emerging needs of AI governance.

Canada

Canada aspires to lead by example with its AI and Data Act, aiming to establish a framework that could serve as a global template for AI regulation. This initiative reflects Canada's commitment to developing responsible AI technologies that align with safety and human rights standards.

China

China has implemented stringent regulations that require AI companies to undergo security reviews before releasing AI models publicly. This regulatory framework is tightly aligned with the country's socio-political values, ensuring that AI development supports national interests and standards.

These regulatory differences create a complex environment for companies operating internationally. Businesses must navigate a patchwork of international laws that can affect everything from AI development to market deployment strategies. The varying regulations not only influence how AI products are designed and what markets they can enter but also shape the global competitive landscape of AI innovation.

For a more detailed exploration of the global AI regulatory landscape, readers can refer to comprehensive reviews and analyses provided by resources from EY [11] and Brookings [8].

6.7 Key Takeaways

- Industrial manufacturers are leading the adoption of AI ethics policies with a high implementation rate of 89% in 2021, indicating a proactive approach to ethical AI across various sectors.
- As AI enhances efficiency and spurs innovation in manufacturing, it brings a responsibility to use AI ethically, ensuring that benefits are equitably distributed and do not exacerbate existing inequalities.
- To implement ethical AI, manufacturers should focus on making AI decisions clear and comprehensible to all stakeholders to build trust and ensure accountability, keeping human intervention in AI processes to correct errors and biases, ensuring AI does not operate autonomously, and regularly updating and auditing AI systems to maintain their effectiveness and unbiased nature over time.
- Ethical AI deployment in manufacturing needs to confront issues like algorithmic biases, cybersecurity risks, and potential job displacements due to automation.

- Public and private investments should focus on AI research that augments human capabilities rather than replaces them, ensuring that AI enhances productivity without sidelining the workforce.
- Regulatory bodies need robust standards for overseeing AI-driven workplaces, ensuring that AI systems comply with national and international laws.
- Organizations that adopt comprehensive strategies for AI governance address AI responsibility concerns more effectively than those without such approaches.
- Effective AI governance in manufacturing should incorporate people, processes, and technology, ensuring that AI applications are responsible, compliant, and beneficial.

Glossary

Algorithmic Fairness The absence of discrimination or bias in automated decision-making systems.

Augmentation (in the context of AI) The use of AI to enhance and support human capabilities rather than replace them.

ECCP (Evaluation of Corporate Compliance Programs) Guidelines issued by the Department of Justice for assessing corporate compliance programs. Website—https://www.justice.gov/criminal/criminal-fraud/page/file/937501/dl

Ethical AI The development and use of artificial intelligence systems that adhere to moral principles and values.

Human Oversight Human monitoring and intervention are involved in AI-driven processes to ensure ethical and responsible operation.

IEEE (Institute of Electrical and Electronics Engineers) A professional association for electronic engineering and electrical engineering. Website: https://www.ieee.org/

Interpretability The degree to which a human can understand the cause of a decision or prediction made by an AI model.

Responsible AI The practice of designing, developing, and deploying AI systems in a manner that empowers people and businesses and fairly impacts customers and society.

Transparency (in AI context) The clarity and openness about how AI systems operate and make decisions.

Watsonx An AI and data platform developed by IBM for building and deploying AI models [12].

References

1. Ethical AI policy implementation by industry 2021 [Internet]. *Statista*. 2021. Access date November 21, 2024. https://www.statista.com/statistics/1232659/worldwide-artificial-intellige nce-implementation-ethics-policies/
2. Inc CD. Exploring ethical AI decision making in manufacturing [Internet]. *Copper Digital*. Access date November 21, 2024. https://copperdigital.com/blog/ethical-implications-ai-dec ision-making-manufacturing/
3. Ryan, M., Antoniou, J., Brooks, L., Jiya, T., Macnish, K., & Stahl, B. (2021). Research and practice of AI ethics: A case study approach Juxtaposing academic discourse with organisational reality. *Science and Engineering Ethics, 27*(2).
4. Case studies [Internet]. Princeton dialogues on AI and ethics. 2018. Access date November 21, 2024. https://aiethics.princeton.edu/case-studies/
5. AI adoption risks relevance in organizations worldwide 2022 | Statista [Internet]. *Statista*. 2022. Access date November 21, 2024. https://www.statista.com/statistics/1381503/ai-adoption-risk-relevance-worldwide/
6. Responsible AI implementations in U.S. 2022 [Internet]. *Statista*. 2022. Access date November 21, 2024. https://www.statista.com/statistics/821592/worldwide-artificial-intelligence-implem entation-in-companies-concerns/
7. Virtue Market Research. AI compliance monitoring market is projected to reach a value of USD 1429.36 Million by 2030 [Internet]. openPR.com. *openPR*; 2023. Access date November 21, 2024. https://www.openpr.com/news/2974386/ai-compliance-monitoring-mar ket-is-projected-to-reach-a-value
8. Engler, A. (2023). The EU and U.S. diverge on AI regulation: A transatlantic comparison and steps to alignment [Internet]. *Brookings*. Access date November 21, 2024. https://www.brookings.edu/articles/the-eu-and-us-diverge-on-ai-regulation-a-transatla ntic-comparison-and-steps-to-alignment/
9. St DCP 655 15th, Washington NS 410, Fax:783-0534 DC 20005 P0. A Look Into the Global AI Regulatory Landscape [Internet]. Disruptive Competition Project. 2024. Access date November 21, 2024. https://project-disco.org/innovation/a-look-into-the-global-ai-regulatory-landscape/
10. Antonini, C. (2023). Navigating the EU AI act: How explainable AI simplifies regulatory compliance [Internet]. *Positive Thinking Company*. Access date November 21, 2024. https://positivethinking.tech/insights/navigating-the-eu-ai-act-how-explainable-ai-simplifies-regulatory-compliance/
11. The Artificial Intelligence (AI) global regulatory landscape: Policy trends and considerations to build confidence in AI. The Artificial Intelligence (AI) global regulatory landscape Policy trends and considerations to build confidence in AI [Internet]. 2024. Access date November 21, 2024. https://assets.ey.com/content/dam/ey-sites/ey-com/en_gl/topics/ai/ey-the-artificial-intelligence-ai-global-regulatory-landscape-v7.pdf
12. IBM. (2024). IBM watsonx | IBM [Internet]. www.ibm.com. Access date November 21, 2024. https://www.ibm.com/watsonx

Chapter 7
The Road Ahead

Abstract The final chapter explores the impact of AI in manufacturing, envisioning an Industry 5.0 era where human intelligence seamlessly collaborates with advanced AI systems. Building upon earlier discussions of AI's current integrations in factory floors, supply chain, and enterprise operations, this chapter discusses the future possibilities and value AI may unlock. It examines how autonomous systems and advanced networking technologies like 5G/6G will revolutionize production ecosystems, enabling real-time decision-making, enhanced safety, and unprecedented customization. The chapter draws parallels with the science fiction film Ex Machina, reflecting on AI's capabilities and challenges in manufacturing. Vision-guided robots, AI-driven 3D printing, and quantum computing illustrate the transformative applications in design, production, and supply chain management. Alongside these advancements, ethical considerations, cybersecurity, and sustainability emerge as critical focus areas, balancing innovation with risk management. Looking ahead to 2031 and beyond, the chapter outlines a vision of fully automated, resilient factories powered by AI. It highlights the importance of proactive adoption strategies, infrastructure modernization, and cultural shifts to utilize AI's full potential. By guiding innovation responsibly, manufacturers can unlock a future of safer, more efficient, and sustainable production systems, marking a paradigm shift in industrial practices.

Keywords Vision-guided robots · Smart factories · AI-driven 3D printing · Predictive maintenance technology · Sustainable manufacturing solutions · 5G-enabled manufacturing

Welcome to the last chapter, exploring how AI will change manufacturing. In earlier chapters, we looked at how AI is already being integrated on factory floors, in supply chain, and across enterprises.

In this chapter, we will discuss the possibilities for where manufacturing with AI may go next. As AI keeps advancing, autonomous systems will likely become more versatile and able to learn, collaborate, and innovate alongside human teammates. As data platforms mature and connect insights to decisions in real-time, AI agents may optimize entire operations end-to-end.

© The Author(s), under exclusive license to Springer Nature Switzerland AG 2025   155
B. Sarkar and R. K. Paul, *AI for Advanced Manufacturing and Industrial Applications*,
https://doi.org/10.1007/978-3-031-86091-1_7

In the future, new kinds of computing could remove limits on today's AI, making it think and intuit more like humans do. These AI assistants would then create new products, designs, and business models free of old constraints. The upcoming wave of progress points to an exciting revolution, though one requiring care and wisdom as we shape how AI rises. Together, these innovations make up Industry 5.0, which combines human and artificial intelligence to improve safety, creativity, and efficiency.

We hope that the ideas and visions shared here provide value and inspiration for the reader's AI manufacturing journey. The brightest innovations, deepest lessons, and toughest trials still await. On we go!

The smart, self-changing factory machines made us think of the robot Ava from the popular movie—Ex Machina. In the movie, a robot named Ava could think and act very much like humans. Now, in real factories, we see robots acting on their own without always being told exactly what to do next.

Drawing from the analogy of how Ava could move and think for herself on the manufacturing floor, we can think of transportation robots that can move materials on their own. In the film, Ava could move around rooms and open doors herself without help, like how flexible automated systems in factories can keep working if something goes wrong. The self-fixing manufacturing lines now keep going even with problems, just like Ava kept moving independently.

Of course, real AI helpers have limits on what they can do on their own, unlike in the movie. But workers today using voice commands with helpful robot partners might feel a little bit like the movie scene with a man watching and talking to Ava through the glass.

Ava made a life-like hand for herself using 3D printing she taught herself. This made us think of how AI now helps engineers print complex factory parts. AI programs create and improve designs customized for 3D printers. By studying past prints, AI guides faster printing with less human effort.

However, much like how Ava ultimately escaped confinement, cybersecurity has emerged as a key priority as intelligent connectivity and automation achieved immense adoption on shop floors. While encryption and access controls aim to secure networked machines, threats persist.

But like Ava broke free at the end of the movie, AI ethics and safety have become very important as factories use more connected AI machines. While there are security protections for smart systems, risks remain. Self-learning AI could also sometimes work against safety if mistakes happen. So, rules to limit AI risks and involve humans in oversight are now critical.

In the movie, Caleb, the human character looking at caged Ava through high-tech glass, is a bit like engineers today with augmented reality dashboards that overlay data on running factory lines. Showing digital performance details on the real equipment improves monitoring. New two-way controls help people guide helpful robot partners using hand gestures, similar to Caleb's conversations with Ava.

While robots in manufacturing have not yet reached the level of sophistication seen in 'Ex Machina's' Ava, the movie highlights important issues regarding AI safety and the potential unease in human-AI interactions. As the film suggests, it is crucial to guide the development of this technology responsibly.

Nonetheless, while factory AI is not as lifelike as the worrying robot in Ex Machina, smart, adaptable machines have now appeared in real manufacturing. The interesting movie seemed to show the future of AI, which is now changing real industries like manufacturing.

7.1   The Vision of AI-Powered Manufacturing in 2031 and Beyond

Having explored the vision for AI in manufacturing and its intriguing parallels with 'Ex Machina's' Ava, let's now turn our attention to the future, beyond the year 2031, to glimpse the full potential of AI. Next, we will provide a summary of the vision for AI in manufacturing in the distant future, and subsequently, we will dive deeper into each topic and the journey in the later sections of Chapter 7.

Figure 7.1 shows how, in 2031 and beyond, manufacturing powered by advanced artificial intelligence will enable fully automated, resilient production ecosystems that can customize output, optimize processes, and revolutionize business models by seamlessly linking the physical and digital.

In 2031 and beyond, industrial manufacturing will experience a dramatic transformation due to AI integration. This revolution, rooted in AI-enabled equipment, enhances production with smart machines that adapt, execute complex tasks precisely, and optimize for efficiency. This leads to significant productivity gains.

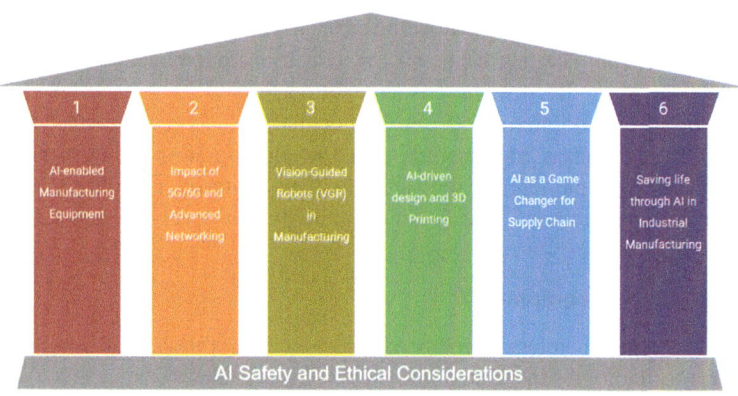

Fig. 7.1 Six (6) Pillars of AI-enablement in manufacturing in 2031 and beyond. *Source* Created by the authors in Canva

The integration of 5G/6G and advanced networking further propels this change. 5G/6G's rapid data transfer enables real-time machine communication, and decentralized networks ensure fluid information flow and instant decision-making.

Vision-guided robots (VGR), equipped with AI vision systems, revolutionize task execution, enhancing precision, safety, and flexibility. AI-driven design coupled with 3D printing accelerates AI adoption in manufacturing. These technologies allow for quick prototyping and efficient production, drastically shortening the development cycle. AI's influence extends to supply chain management, analyzing trends, forecasting demand, and optimizing inventory, thus streamlining operations and meeting customer needs more effectively. A notable impact of AI is the enhancement of safety measures. Real-time monitoring and predictive systems make factories safer and more humane by performing high-risk tasks and preventing accidents.

In the future, AI's role in manufacturing will demonstrate how it can redefine industries, improve safety, and drive innovation. The journey illustrates an expansive future potential for AI in further evolving manufacturing from the present day.

Now, let's explore the aspects discussed in this chapter.

7.2   AI-Enabled Manufacturing Equipment

Industrial machinery embedded with artificial intelligence software and sensors for data collection and analytics is becoming the new standard for optimizing manufacturing operations. Intelligent capabilities like predictive maintenance scheduling and cognitive quality control are driving increased adoption across factory floors and supply chain.

The utilization of AI-driven predictive maintenance has evolved from preventing costly breakdowns to orchestrating maintenance schedules to optimize production. The vendor landscape will likely mature the most, and more solutions will be available to meet customer complexities and needs. As industrial equipment companies switch to the servitization model, it will act as an additional driver.

Additionally, the adoption of AI-powered cognitive analytics to enhance quality control is likely to grow and mature with the incorporation of Gen-AI. Next-generation artificial intelligence will amplify data analysis capabilities, bringing product defect detection and process adjustments to new levels of automation and precision.

What is AI-enabled equipment?

AI-enabled equipment refers to industrial machinery and tools that have artificial intelligence technologies integrated into them. This includes:

- Smart sensors, controllers, and connectivity that allow real-time data collection
- Embedded AI chips or edge processors that run algorithms locally
- Advanced software capabilities like machine learning, computer vision, and natural language processing

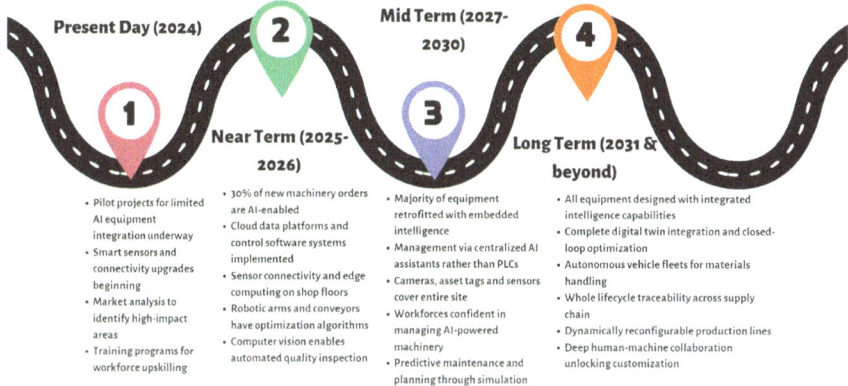

Fig. 7.2 Journey for adopting AI-enabled equipment for smart Manufacturing. *Source* Created by the authors in Canva

- Human–machine interfaces with augmented reality for insights and alerts.

  Key examples of AI functionalities in equipment include:

- Robotic arms automatically optimize speed and precision
- Power tools self-adapting torque and monitoring wear
- Conveyor systems dynamically adjust routes to reduce bottlenecks
- Anomaly detection spotting early signs of faults
- Computer vision for automated quality inspection.

Timeline for adopting AI-enabled equipment:

Figure 7.2 describes from present-day co-bot prototyping and simulation to mid-term scaled production of customizable AI-powered collaborative robots and, ultimately, the retirement of legacy robots in favor of flexible, self-learning AI-enabled equipment or advanced co-bots seamlessly co-creating complex customized orders with humans beyond 2031.

7.3 Impact of 5G/6G and Advanced Networking

The manufacturing industry is in the midst of a technological revolution driven by smart manufacturing leveraging IoT, AI, and ML. This evolution significantly improves efficiency, real-time monitoring, and process control.

5G (fifth generation) or 6G (sixth generation) is the next-generation wireless standard that goes beyond cellular mobile communications to provide an overarching enhanced connectivity ecosystem spanning high-speed broadband access, massive machine-type communications, and ultra-reliable low-latency links for mission-critical applications.

While previous networks supported this shift, the advent of 5G and then 6G has the potential to supercharge the revolution, offering high-speed, low-latency, and large-bandwidth connectivity. Many businesses have already embraced 5G to implement Industry 4.0 initiatives, maximizing the benefits of advanced technologies.

Table 7.1 depicts the timeline for an ambitious adoption journey as manufacturers rapidly deploy 5G for competitive advantage. Initial trials give way to major pilots, and then production-level deployments through 2024 as benefits become clear. By 2025, 5G will become a mature operational component, sparking an ongoing innovation cycle to apply 5G's capabilities. This table provides a high-level overview of the expected progression if the current momentum is sustained.

5G connectivity is crucial for enabling wider adoption of artificial intelligence in the manufacturing industry for several key reasons:

1. High-speed, low-latency 5G allows real-time collection and analysis of data from sensors and equipment on factory floors. This powers AI applications like predictive maintenance, quality optimization, and intelligent automation that need timely data.
2. As industrial machinery gets embedded with more sensors and controllers, 5G provides a reliable high-bandwidth wireless infrastructure to handle massive flows of sensor data. Legacy networks would fail to attempt the same scale.
3. 5G facilitates advanced concepts like digital twins and digital threads, which require the seamless connectivity of the physical manufacturing environment with cloud-based simulations and AI models for monitoring, analytics, etc.
4. For human workers, 5G allows the use of AR/VR tools to implement AI-assisted manufacturing techniques by overcoming latency bottlenecks faced in assistive technologies today over WiFi or 4G.
5. 5G enables decentralized AI processing to happen at the edge near machinery inside plants via multi-access edge computing instead of sending all data to distant clouds. This is vital for real-time applications.

Table 7.1 Strategies for managing data challenges in AI-driven manufacturing

Year	Adoption stage	Description
2024	Majority adoption	• Widespread production deployments on factory floors and across supply chain • Furthermore, 76% of manufacturers have adopted 5G in some form, per the 2021 VentreBeat report
2025	Mature integration	• 5G is fully integrated into core manufacturing operations, analytics, and IoT systems • Legacy upgrades completed for comprehensive adoption
2026 and beyond	Optimization and innovation	• The focus shifts to maximizing 5G-enabled use cases and the adoption of 6G • Creative new applications continue expanding capabilities

In essence, AI and 5G will combine to profoundly automate and optimize nearly all aspects of manufacturing through ubiquitous intelligent connectivity. 5G provides a standardized high-performance infrastructure, allowing manufacturing companies to tap into the full range of efficiency and quality gains promised by Industry 4.0.

7.3.1   From Hierarchical to Advanced Networks

Legacy manufacturing factories followed centralized, hierarchical models with limited flexibility. Future smart factories will instead have distributed, self-optimizing network architectures enabled by AI.

What can Advanced Networks with AI/ML capabilities do?

By leveraging AI/ML and advanced networking in tandem, next-generation smart manufacturing distributed networks powered by the 5G combination discussed earlier enable a variety of powerful capabilities:

- Localized ML algorithms balancing control tasks across nodes
- Edge devices like machine sensors running adaptive analytics
- Peer-to-peer coordination communicating constraints
- Backup neural networks for fail-safe redundancy
- Cloud-managed meta-learning continually enhances models
- Control loops dynamically reconfiguring based on insights.

This shift brings extreme resilience, scalability, and autonomous optimization. Human roles transition from control programming to training oversight.

Advanced Network powered by 5G with AI/ML Capabilities:

Figure 7.3 describes the overview of how the adoption of AI/ML will transform industrial network architectures and a potential timeline.

6G networks will enable very fast connections in future manufacturing ecosystems. The super-fast connectivity will link real factory equipment and advanced digital twin simulations. Instant sharing of data at a massive scale is made possible by 6G communication platforms. It will allow humans and AI systems to collaborate via virtual or augmented reality interfaces. Together, they can invent creative production solutions and customize output to individual buyer needs. Seamless integration of physical and digital operations, driven by AI-human teamwork suited for humans, will be at the core of Industry 5.0. By efficiently synchronizing both worlds, 6G substrates facilitate that emerging paradigm.

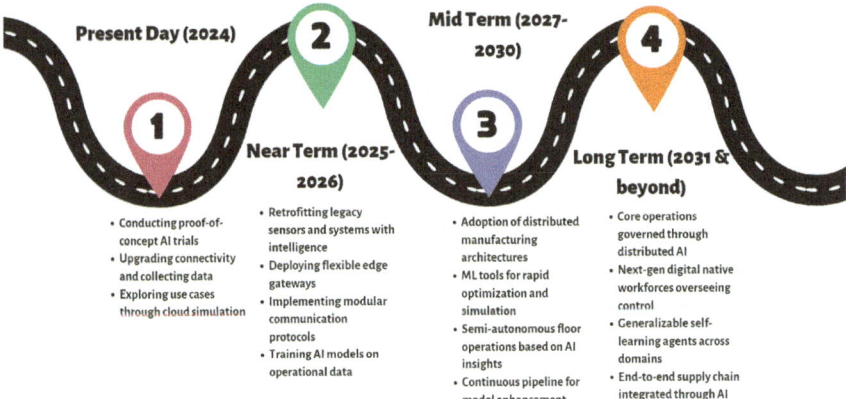

Fig. 7.3 Evolution of industrial network architectures. *Source* Created by the authors in Canva

7.4 Vision-Guided Robots (VGR) in Manufacturing

The manufacturing industry is on the brink of significant disruption with the advent of Generalized Robotics, specifically Vision-Guided Robots (VGR). These robots, equipped with advanced vision sensors and sophisticated image-processing algorithms, are designed to perceive, interpret, and interact with their environment. The integration of VGR with AI and Machine Learning (ML) technologies has exponentially increased their capabilities, allowing them to perform tasks with a level of precision and adaptability previously unattainable.

What will Next-Generation Generalized Robots be Capable of in Manufacturing?

The primary aim of these next-generation robots is to augment human capabilities, providing a level of precision, efficiency, and adaptability in manufacturing processes that were previously unachievable. This leap forward in robotics will not only improve product quality and consistency but also revolutionize the way manufacturing operations are conducted. By shouldering tasks that are either too hazardous, intricate, or monotonous for humans, these robots will facilitate a safer, more productive, and innovative manufacturing environment.

The flexible, intelligent robots of the future will:

- Coordinate across a wide variety of components and sub-assemblies across product families
- Smoothly collaborate in shared factory workspaces alongside operators
- Continuously expand manufacturing skills through AI and observational learning
- Understand verbal commands and factory floor contextual cues
- Perform non-repetitive tasks like equipment maintenance and repair
- Dynamically switch roles between product variants with minimal reconfiguration
- Make autonomous decisions by processing data at the edge

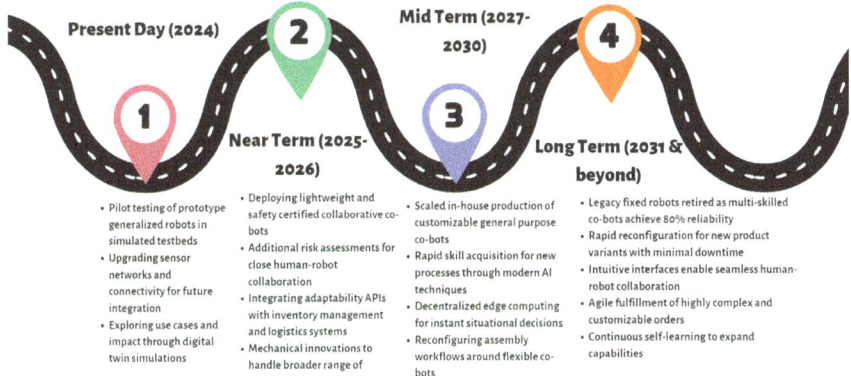

Present Day (2024)

Near Term (2025-2026)

Mid Term (2027-2030)

Long Term (2031 & beyond)

- Pilot testing of prototype generalized robots in simulated testbeds
- Upgrading sensor networks and connectivity for future integration
- Exploring use cases and impact through digital twin simulations

- Deploying lightweight and safety certified collaborative co-bots
- Additional risk assessments for close human-robot collaboration
- Integrating adaptability APIs with inventory management and logistics systems
- Mechanical innovations to handle broader range of components

- Scaled in-house production of customizable general purpose co-bots
- Rapid skill acquisition for new processes through modern AI techniques
- Decentralized edge computing for instant situational decisions
- Reconfiguring assembly workflows around flexible co-bots

- Legacy fixed robots retired as multi-skilled co-bots achieve 80% reliability
- Rapid reconfiguration for new product variants with minimal downtime
- Intuitive interfaces enable seamless human-robot collaboration
- Agile fulfillment of highly complex and customizable orders
- Continuous self-learning to expand capabilities

Fig. 7.4 Journey for the adoption of generalized robotics in manufacturing. *Source* Created by the authors in Canva

- Communicate dynamic production needs and constraints with workers and machinery.

These adaptable multi-purpose machines will amplify human capabilities across diverse factory situations.

Expected Timeline for Generalized Robotics Adoption in Manufacturing:

Figure 7.4 describes present-day co-bot prototyping and simulation to mid-term scaled production of customizable AI-powered collaborative robots and, ultimately, the retirement of legacy robots in favor of flexible, self-learning co-bots seamlessly co-creating complex customized orders with humans beyond 2031.

7.5   AI-Driven Design and 3D Printing

The integration of artificial intelligence (AI) and 3D printing technology is unlocking major innovations in manufacturing processes and final products. As demand rises for custom and personalized offerings, particularly in sectors like healthcare, AI's design enhancement capabilities are proving crucial—augmenting limited designer bandwidth. By combining AI's predictive analytics with 3D printing's flexible on-demand production, manufacturers can optimize workflows, reduce errors, and deliver tailored solutions at scale.

This symbiotic relationship manifests through two key mechanisms:

AI-driven Product Design

Sophisticated generative AI can autonomously create novel product designs tailored to consumer preferences and emerging market trends. Rather than simply enhancing human-generated concepts, advanced algorithms drive the creative process itself—significantly expediting iteration cycles and amplifying design inventiveness.

AI-optimized 3D Printing Workflow

AI predictive analytics, simulation, and monitoring tools allow streamlined design-to-print transitions. By spotting potential printability issues early and optimizing build parameters for specific 3D printer setups, manufacturers can maximize throughput and part quality. These AI-enabled optimizations also support the rapid incorporation of design changes and customizations.

What is the synergy between AI-driven Product Design and 3D Printing?

Innovative AI-driven product design refers to the use of AI on the front-end to autonomously generate innovative industrial product designs and concepts based on data analysis. It acts as an automated creative design assistant.

Meanwhile, 3D Printing (an Additive Manufacturing technology) applies AI tools to the back-end 3D printing process to smooth workflows and optimize builds. Predicting and adjusting for issues during job preparation, simulation, and printing enables optimal digital manufacturing on demand.

Together, the creative potential of generative AI and the responsiveness of 3D printing promise a new era of nimble, hyper-customized production. The manufacturing sector's digital transformation continues to accelerate thanks to these mutually reinforcing technologies.

The market size of the 3D printing market was valued at USD 17.4 Billion in 2022 and is expected to grow to 37.2 billion USD in 2026. This demonstrates a projected compound annual growth rate (CAGR) of 16% [1].

The synergy between AI and 3D Printing is transforming manufacturing and design processes. This integration is crucial across various industries, particularly those requiring high levels of customization and precision, like healthcare and aerospace.

7.6 AI as a Game Changer for Supply Chain

Inventory management enabled by artificial intelligence promises a revolution in smart manufacturing and retail operations. AI-driven multi-channel planning and optimization will be the pillar transforming enterprise supply chain performance this decade. By dynamically orchestrating interconnected stock levels with precision, companies can achieve unprecedented efficiency, resilience, and sustainability.

Figure 7.5 shows how, by perpetually balancing priorities amidst variability, AI-optimized inventory orchestration promises to realize profit-maximizing, waste-minimizing, and customer-delighting agile manufacturing enabled through data-driven digital resilience.

The specifics of how AI can shape future inventory management and reshape smart manufacturing include

Fig. 7.5   AI-powered cycle of inventory management. *Source* Created by the authors in Canva

Better Demand Planning and Re-stocking

AI systems will leverage extensive real-time and historical data encompassing sales, pricing, events, and more to predict inventory requirements. By always learning from new signals, the AI can automatically set and adjust stock levels in different locations to match fulfillment patterns. It is more advanced than old statistical methods.

More Accurate Delivery Estimates

The AI software combines data on logistics, weather issues, etc., to calculate precise delivery time estimates for customers' orders. Retailers can set rules on how to balance cost and customer service per channel using AI-powered insights. Engineers can check exactly how the AI made each ETA decision to keep improving it.

Dynamic Safety Stock Tailored by AI

Instead of having fixed blanket buffer stock levels, AI algorithms will simulate likely demand changes from seasonality or random events. Then, the AI can tune the safety reserves in each place based on volatility patterns to minimize costs while still meeting service targets. The AI also automatically adapts its inventory policies using real results to handle new uncertainties.

7.7 Saving Life Through AI in Industrial Manufacturing

We have discussed below how AI can be a hero and a potential villain for compromising cybersecurity systems in the manufacturing enterprises of the future.

Advanced AI systems are poised to revolutionize manufacturing, acting as heroic problem-solvers that protect workers and optimize operations. As AI adoption accelerates, however, the very cybersecurity measures intended to safeguard manufacturers may prove to be an unexpected nemesis that stifles progress.

Cutting-edge AI solutions promise enhanced efficiency, quality, and safety, anticipating risks and recommending actions with superhuman speed and precision. Like a manufacturing guardian angel, AI can guide workers away from hazards in real-time while coordinating predictive maintenance and adjustments across machinery.

However, most AI systems rely on widespread connectivity and data aggregation to derive such valuable insights. This extensive integration exposes inherent vulnerabilities that cybercriminals can exploit. Though cybersecurity controls aim to preempt attacks, overly rigid constraints may prevent AI from reaching its full potential.

As manufacturers rush to implement AI, security teams exert cautious restraint, limiting access and connectivity to mitigate threats. However, compartmentalization and strict permissions make gathering, processing, and acting upon the holistic data that AI depends on nearly impossible.

Without unfettered machine-speed access to operational datasets, AI cannot fully showcase its game-changing capabilities. What initially appeared to be a heroic AI revolution stands to be undermined by well-intentioned but short-sighted cybersecurity protocols.

Manufacturers must thoughtfully balance enabling AI as a creative force for good against neutralizing risks. By taking an ambitious approach and proactively strengthening defenses around AI integration points rather than reactively limiting functionality, they can nurture AI's heroism while controlling its vulnerability. With careful nurturing, AI can derive a new era of safer, more efficient, and resilient production.

7.8 Potential Impact of Quantum Computing on AI in Manufacturing

Quantum computing is an exciting new technology that could revolutionize how we process information and solve complex problems. Unlike traditional computers that use bits (0s and 1s), quantum computers use quantum bits or qubits. These qubits can exist in multiple states at once, allowing quantum computers to perform certain calculations much faster than regular computers [2].

In manufacturing, this speed boost could have a huge impact on AI applications. Quantum computers could run incredibly detailed simulations of manufacturing processes, materials, and product designs. This would allow companies to test and optimize their operations virtually, saving time and resources. For example, a car manufacturer could use quantum-powered AI to simulate crash tests or aerodynamics with unprecedented accuracy [3].

Quantum computing could supercharge machine learning algorithms, allowing AI systems to learn and adapt much more quickly. This could lead to more intelligent and responsive manufacturing systems. For instance, an AI system could analyze vast amounts of production data in real-time, instantly adjusting processes to maintain optimal quality and efficiency [4].

Many manufacturing challenges involve complex optimization problems, like scheduling production or managing supply chain. Quantum computers are particularly good at solving these types of problems. They could help AI systems find the best solutions much faster, leading to more efficient operations and cost savings [5].

Quantum computing could accelerate the discovery of new materials with specific properties. This could revolutionize product design and manufacturing processes. AI systems powered by quantum computers could simulate countless molecular combinations, potentially leading to stronger, lighter, or more sustainable materials [6].

Quantum computing offers new possibilities for encryption, which could help secure sensitive manufacturing data and AI systems against cyber threats. This is particularly important as factories become more connected and reliant on data [7].

By processing more data and variables simultaneously, quantum-enhanced AI could predict equipment failures with greater accuracy. This could help prevent costly breakdowns and extend the life of manufacturing equipment.

Some manufacturing challenges are so complex that they're practically impossible to solve with current technology. Quantum computing could change that, allowing AI to tackle problems we can't even approach today.

However, it's important to note that practical, large-scale quantum computers are still years away. Many technical challenges need to be overcome before we can fully harness their power. Additionally, not all problems will benefit from quantum computing—in many cases, traditional computers will still be more practical and cost-effective [8].

As quantum technology develops, manufacturers will need to stay informed and prepare for this shift. This might involve partnering with quantum computing researchers, experimenting with quantum-inspired algorithms on classical computers, or identifying areas where quantum computing could have the biggest impact on their operations [9].

In conclusion, while quantum computing is still in its early stages, its potential to enhance AI in manufacturing is enormous. As this technology matures, it could lead to smarter factories, better products, and entirely new ways of approaching manufacturing challenges.

7.9   Sustainability, Carbon Footprint and AI in Manufacturing

Artificial Intelligence (AI) is becoming a powerful tool in the fight against climate change and environmental degradation. In manufacturing, AI can help companies reduce their carbon footprint, minimize waste, and create more sustainable processes. Here's how AI is driving sustainability initiatives in manufacturing:

AI systems can analyze energy usage patterns in factories and optimize them in real time. For example, AI can adjust lighting, heating, and cooling based on occupancy and production schedules. It can also identify energy-hungry machines and suggest more efficient alternatives or operating methods.

AI can help minimize waste in several ways. Computer vision systems can spot defects early in the production process, reducing the amount of material that becomes waste. AI can also optimize cutting patterns for materials like fabric or metal, maximizing the use of raw materials. Furthermore, AI can analyze waste streams and suggest ways to recycle or repurpose materials that would otherwise be discarded [10].

By predicting when machines are likely to fail, AI helps prevent unexpected breakdowns. This not only saves on repair costs but also reduces waste from defective products and extends the lifespan of equipment, reducing the need for replacements and the associated environmental impact [11].

AI can analyze complex supply chain data to find more efficient routes, reduce empty truck miles, and optimize inventory levels. This leads to lower fuel consumption and less waste from overstocking or spoilage. For instance, Walmart used AI to optimize its truck routes, saving millions of miles driven and tons of CO_2 emissions annually [12].

AI can assist in designing products that are more environmentally friendly. It can suggest alternative materials, simulate product lifecycles to improve durability, and even design products for easier recycling at the end of their life [13].

AI systems can provide real-time monitoring of emissions from manufacturing processes. They can suggest adjustments to reduce emissions and help companies comply with environmental regulations more effectively [14].

AI can assist in collecting and analyzing sustainability data, making it easier for companies to report on their environmental impact and set improvement goals [15].

While AI offers many opportunities for improving sustainability in manufacturing, it's important to note that AI systems themselves consume energy. Companies need to balance the environmental benefits of AI against its energy use.

Moreover, successfully implementing AI for sustainability requires a commitment from company leadership, investment in technology and skills, and often a cultural shift towards prioritizing environmental concerns.

In conclusion, AI is proving to be a valuable ally in making manufacturing more sustainable. As AI technology continues to advance, we can expect even more innovative applications that help reduce the environmental impact of industrial processes. By embracing these technologies, manufacturers can not only reduce their carbon footprint and waste but also often realize cost savings and improved efficiency, making sustainability initiatives a win–win for business and the environment.

7.10 Key Takeaways

- The adoption of intelligent machinery and tools with embedded AI will lead to significant improvements in productivity and efficiency. Predictive maintenance and cognitive quality control will become standard features, enhancing the longevity and performance of manufacturing equipment.
- The integration of 5G and 6G technologies will revolutionize manufacturing by enabling high-speed, low-latency, and reliable communication, essential for real-time operations. These networks will support more decentralized, intelligent decision-making processes and facilitate advanced manufacturing practices like digital twins.
- VGRs will transform manufacturing floors with their precision and flexibility, allowing for more complex and varied tasks to be automated. The synergy between AI and robotics will lead to safer and more efficient operations, minimizing human involvement in hazardous environments.
- AI will enhance the capabilities of 3D printing by optimizing design processes and print configurations, leading to faster production times and higher-quality products. This combination will enable manufacturers to respond more swiftly to market demands with customized and complex products.
- AI will transform supply chain management by improving demand forecasting, inventory management, and logistical efficiency. These improvements will lead to reduced waste, lower costs, and enhanced ability to adapt to market changes and consumer demands.
- AI's role in improving safety protocols through predictive analytics and real-time monitoring will make factories safer and reduce workplace accidents. Ethical considerations and cybersecurity will be paramount as reliance on AI increases, necessitating robust security measures and ethical guidelines to govern AI usage.
- The economic landscape of manufacturing will shift dramatically, with AI driving down costs, improving operational efficiency, and opening new markets through innovative products and services. Companies that proactively integrate AI will gain a competitive edge, setting industry standards and leading market trends.
- Companies must develop strategic frameworks to assess and integrate AI technologies, balancing potential benefits against costs and risks. Investment in training and development will be crucial to equip employees with the skills needed to operate in an AI-enhanced manufacturing environment.
- The integration of AI will require cultural adjustments within organizations, promoting collaboration between humans and machines. Change management practices will be vital to ensure smooth transitions and acceptance of new technologies among the workforce.

- Continuous innovation and adaptation will be necessary to keep pace with technological advancements and evolving market demands. Establishing iterative processes for technology evaluation and implementation will help companies stay ahead in a rapidly changing environment.

7.11 Thank You for Embarking on the Journey of Exploring Manufacturing with AI

The innovations discussed signal the enormous disruptions artificial intelligence will soon spark. As the pace of technological advancement accelerates into the 2030s, companies have a narrow window to adapt before losing competitiveness. Crucial imperatives stand out:

First, organizations require frameworks to guide AI investment tradeoffs even with constrained budgets. Evaluating use cases via rapid small-scale pilots with defined performance metrics determines viability prior to full deployments.

Second, modernizing surrounding infrastructure becomes critical so AI can enhance rather than overwhelm existing systems. Upgrading connectivity, data flows, and simulation platforms provide the resilient, responsive foundations needed to unlock automation benefits.

Finally, culture change management must also commence as AI compels evolving skill sets and decision hierarchies. From shop floors to board rooms, integrating human–machine collaboration in creative thinking, ethics oversight, and risk assurance is mandatory.

While monumental in scope, first-movers stand to attain exponential advantage as AI adoption inflects. Prioritizing maturity initiatives today future-proofs manufacturing operations for maximal competitiveness tomorrow when AI proliferation gathers unstoppable momentum. With vision, rigor, and sustained experiments, the transformation journey beckons.

We thank you deeply for joining us on this exploration of possibilities as artificial intelligence promises to reshape manufacturing in the years ahead.

Glossary

6G The sixth generation of wireless technology, expected to succeed 5G with even higher speeds and lower latency.

Additive Manufacturing A manufacturing process that creates objects by adding material layer by layer, often synonymous with 3D printing.

Augmented Reality (AR) Technology that superimposes digital information on the user's view of the real world.

Digital Twin A virtual representation of a physical object or system used to simulate and analyze its performance.

Edge Computing Processing data near the source of data generation, at the "edge" of the network, rather than in a centralized data-processing warehouse.

Generalized Robotics Robots that are designed to perform a wide variety of tasks and adapt to different environments, as opposed to specialized robots.

Industry 5.0 The next phase of the industrial revolution focuses on the collaboration between humans and smart systems.

Multi-access Edge Computing (MEC) A network architecture concept that enables cloud computing capabilities at the edge of the cellular network.

Servitization A business model where companies offer services as an integral part of their products.

Vision-Guided Robots (VGR) Robots equipped with vision systems that allow them to see and interpret their environment.

References

1. Statista. (2021). Global 3D printing industry market size [Internet]. *Statista.* Access date November 21, 2024. https://www.statista.com/statistics/315386/global-market-for-3d-pri nters/
2. Schneider, J., & Smalley, I. (2024). What is quantum computing? [Internet]. *IBM.* Access date November 21, 2024. https://www.ibm.com/topics/quantum-computing
3. TDM Systems provides digital power for small manufacturers | *Manufacturing Tomorrow* [Internet]. Manufacturingtomorrow.com. Access date November 21, 2024. https://www.man ufacturingtomorrow.com/article/2021/03/quantum-computing-in-manufacturing/16831
4. Biamonte, J., Wittek, P., Pancotti, N., Rebentrost, P., Wiebe, N., & Lloyd, S. (2017). Quantum machine learning. *Nature, 549*(7671), 195–202. https://doi.org/10.1038/nature23474
5. Quantum computing use cases are getting real—What you need to know [Internet]. McKinsey & Company. *McKinsey & Company*; 2021. Access date November 21, 2024. https://www.mck insey.com/industries/industrials-and-electronics/our-insights/quantum-computing-use-cases-are-getting-real-what-you-need-to-know
6. Pyzer-Knapp, E. O., Pitera, J. W., Staar, P. W. J., Takeda, S., Laino, T., Sanders, D. P., Sexton, J., Smith, J. R., & Curioni, A. (2022). Accelerating materials discovery using artificial intelligence, high performance computing and robotics. *NPJ Computational Materials, 8*(1), 1–9. https://doi.org/10.1038/s41524-022-00765-z
7. Post-quantum cryptography. *NIST* [Internet]. Access date November 21, 2024. https://www.nist.gov/programs-projects/post-quantum-cryptography
8. Patrick, C. S. P. (2019). The problem with quantum computers [Internet]. *Scientific American.* Access date November 21, 2024. https://www.scientificamerican.com/blog/observations/the-problem-with-quantum-computers/
9. A business leader's guide to quantum technology [Internet]. *Deloitte Insights.* Access date November 21, 2024. https://www2.deloitte.com/us/en/insights/topics/innovation/quantum-computing-business-applications.html
10. Olawade, D. B., Oluwaseun Fapohunda, Wada, O. Z., Usman, S. O., Ige, A. O., Ajisafe, O., & Oladapo, B. I. (2024). Smart waste management: A paradigm shift enabled by artificial intelligence. *Waste Management Bulletin, 2*(2). https://doi.org/10.1016/j.wmb.2024.05.001
11. IBM. (2023). What is predictive maintenance? | *IBM* [Internet]. www.ibm.com. Access date November 21, 2024. https://www.ibm.com/topics/predictive-maintenance

12. Walmart. (2024). Walmart commerce technologies launches AI-powered logistics product [Internet]. corporate.walmart.com. Access date November 21, 2024. https://corporate.wal mart.com/news/2024/03/14/walmart-commerce-technologies-launches-ai-powered-logistics-product

13. Sustainability starts in the design process, and AI can help [Internet]. *MIT Technology Review*. 2022. Access date November 21, 2024. https://www.technologyreview.com/2022/01/19/104 3819/sustainability-starts-in-the-design-process-and-ai-can-help/

14. Degot, C., Duranton, S., Frédeau, M., Hutchinson, R. (2021). Reduce carbon and costs with the power of AI [Internet]. *BCG Global*. Access date November 21, 2024. https://www.bcg.com/publications/2021/ai-to-reduce-carbon-emissions

15. Sustainability solutions, software & services [Internet]. *SAP*. Access date November 21, 2024. https://www.sap.com/products/sustainability.html

Index

The manufacturer's authorised representative in the EU is Springer
Nature Customer Service Centre GmbH, Europaplatz 3, 69115 Heidelberg,
Germany. If you have any concerns regarding our products, please
contact ProductSafety@springernature.com

Printed and bound by CPI Group (UK) Ltd, Croydon, CR0 4YY
24/04/2026
02096316-0001